T0207263

Computational Biology

Volume 21

The *Computational Biology* series publishes the very latest, high-quality research devoted to specific issues in computer-assisted analysis of biological data. The main emphasis is on current scientific developments and innovative techniques in computational biology (bioinformatics), bringing to light methods from mathematics, statistics and computer science that directly address biological problems currently under investigation.

The series offers publications that present the state-of-the-art regarding the problems in question; show computational biology/bioinformatics methods at work; and finally discuss anticipated demands regarding developments in future methodology. Titles can range from focused monographs, to undergraduate and graduate textbooks, and professional text/reference works.

More information about this series at http://www.springer.com/series/5769

Jeremy Ramsden

Bioinformatics

An Introduction

Third Edition

 Springer

Jeremy Ramsden
The University of Buckingham
Buckingham
UK

ISSN 1568-2684
Computational Biology
ISBN 978-1-4471-6865-2   ISBN 978-1-4471-6702-0   (eBook)
DOI 10.1007/978-1-4471-6702-0

Springer London Heidelberg New York Dordrecht
© Springer-Verlag London 2015
Softcover reprint of the hardcover 3rd edition 2015
1st edition: © Kluwer Academic Publishers 2004
2nd edition: © Springer-Verlag London 2009

Printed on acid-free paper

Springer-Verlag London Ltd. is part of Springer Science+Business Media (www.springer.com)

Mi a tudvágyat szakhoz nem kötők,
Átpillantását vágyuk az egésznek

Imre Madách

Preface to the Third Edition

The publication of this third edition has provided the opportunity to carefully scrutinize the entire contents and update them wherever necessary. Overview and aims, organization and features, and target audiences remain unchanged. The main additions are in Part III (Applications), which has acquired new sections or chapters on the seemingly ever-expanding "omics"—now metagenomics, toxicogenomics, glycomics, lipidomics, microbiomics, and phenomics are all covered, albeit mostly briefly. The increasing involvement of information theory with ecosystems management, which is undoubtedly a part of biology, was felt to warrant a new chapter on that topic. The nervous system has also been explicitly included: it is indubitably an information processor and at the same time biological and, therefore, certainly warrants inclusion, although consideration of the vastness of the topic and its extensive coverage elsewhere has kept the corresponding chapter brief. A section on the automation of biological research now concludes the work.

In his contribution, entitled "The domain of information theory in biology," to the 1956 *Symposium on Information Theory in Biology*,[1] Henry Quastler remarks (p. 188) that "every kind of structure and every kind of process has its informational aspect and can be associated with information functions. In this sense, the domain of information theory is universal—that is, information analysis *can* be applied to absolutely anything." This sentiment continues to pervade the present work.

The author takes this opportunity to thank all those who kindly commented on the second edition.

January 2015

---

[1]Yockey.

Preface to the Second Edition

Overview and Aims

This book is intended as a self-contained guide to the entire field of bioinformatics, interpreted as the application of information science to biology. There is a strong underlying belief that information is a profound concept underlying biology, and familiarity with the concepts of information should make it possible to gain many important new insights into biology. In other words, the vision underpinning this book goes beyond the narrow interpretation of bioinformatics sometimes encountered, which may confine itself to specific tasks such as the attempted identification of genes in a DNA sequence.

Organization and Features

The chapters are grouped into three parts, respectively covering the relevant fundamentals of information science, overviewing all of biology, and surveying applications. Thus Part I (Fundamentals) carefully explains what information is, and discusses attributes such as value and quality, and its multiple meanings of accuracy, meaning, and effect. The transmission of information through channels is described. Brief summaries of the necessary elements of set theory, combinatorics, probability, likelihood, clustering, and pattern recognition are given. Concepts such as randomness, complexity, systems, and networks, needed for the understanding of biological organization, are also discussed. Part II (Biology) covers both organismal (ontogeny and phylogeny, as well as genome structure) and molecular aspects. Part III (Applications) is devoted to the most important practical applications of bioinformatics, notably gene identification, transcriptomics, proteomics, interactomics (dealing with networks of interactions), and metabolomics. These chapters start with a discussion of the experimental aspects (such as DNA sequencing in the genomics chapter), and then move on to a thorough discussion of how the data are analysed. Specifically, medical applications are grouped in a separate chapter.

A number of problems are suggested, many of which are open-ended and intended to stimulate further thinking. The bibliography points to specialized monographs and review articles expanding on material in the text, and includes guide references to very recently reported research not yet to be found in reviews.

Target Audiences

This book is primarily intended as a textbook for undergraduates, for whom it aims to be a complete study companion. As such, it will also be useful to the beginning graduate student.

A secondary audience is physical scientists seeking a comprehensive but succinct guide to biology, and biological scientists wishing to better acquaint themselves with some of the physicochemical and mathematical aspects that underpin the applications.

It is hoped that all readers will find that even familiar material is presented with fresh insight, and will be inspired to new thoughts.

The author takes this opportunity to thank all those who gave him their comments on the first edition.

May 2008

Preface to the First Edition

This little book attempts to give a self-contained account of bioinformatics, so that the newcomer to the field may, whatever his point of departure, gain a rather complete overview. At the same time it makes no claim to be comprehensive: The field is already too vast—and let it be remembered that although its recognition as a distinct discipline (i.e., one after which departments and university chairs are named) is recent, its roots go back a long time.

Given that many of the newcomers arrive from either biology or informatics, it was an obvious consideration that for the book to achieve its aim of completeness, large portions would have to deal with matter already known to those with backgrounds in either of those two fields; that is, in the particular chapters dealing with them, the book would provide no information for them. Since such chapters could hardly be omitted, I have tried to consider such matter in the light of bioinformatics as a whole, so that even the student ostensibly familiar with it could benefit from a fresh viewpoint.

In one regard especially, this book cannot be comprehensive. The field is developing extraordinarily rapidly and it would have been artificial and arbitrary to take a snapshot of the details of contemporary research. Hence I have tried to focus on a thorough grounding of concepts, which will enable the student not only to understand contemporary work but should also serve as a springboard for his or her own discoveries. Much of the raw material of bioinformatics is open and accessible to all via the Internet, powerful computing facilities are ubiquitous, and we may be confident that vast tracts of the field lie yet uncultivated. This accessibility extends to the literature: Research papers on any topic can usually be found rapidly by an Internet search and, therefore, I have not aimed at providing a comprehensive bibliography.

In bioinformatics, so much is to be done, the raw material to hand is already so vast and vastly increasing, and the problems to be solved are so important (perhaps the most important of any science at present), we may be entering an era comparable to the great flowering of quantum mechanics in the first three decades of the twentieth century, during which there were periods when practically every doctoral thesis was a major breakthrough. If this book is able to inspire the student to take up some of the challenges, then it will have accomplished a large part of what it sets out to do.

Indeed, I would go further to remark that I believe that there are still comparatively simple things to be discovered and that many of the present directions of work in the field may turn out not to be right. Hence, at this stage in its development the most important thing is to facilitate that viewpoint that will facilitate new discoveries. This belief also underlies the somewhat more detailed coverage of the biological processes in which information processing in nature is embodied than might be considered customary.

A work of this nature depends on a long history of interactions, discussions, and correspondence with many present and erstwhile friends and colleagues, some of whom, sadly, are no longer alive. I have tried to reflect some of this debt in the citations. Furthermore, many scientific subjects and methods other than those mentioned in the text had to be explored before the ones best suited to the purpose of this work could be selected, and my thanks are due to all those who helped in these preliminary studies. I should like to add a special word of thanks to Victoria Kechekhmadze for having so ably drawn the figures.

January 2004

Contents

Introduction

Information is central to life. The principle enunciated by Crick, that information flows from the gene (DNA) to the protein, occupies such a key place in modern molecular biology that it is frequently referred to as the "central dogma": DNA acts as a template to replicate itself, DNA is transcribed into RNA, and RNA is translated into protein.

The mission of biology is to answer the question "What is life?" For many centuries, the study of the living world proceeded by examination of its external characteristics (i.e., of phenotype, including behaviour). This led to Linnaeus' hierarchical classification. A key advance was made about 150 years ago when Mendel established the notion of an unseen heritable principle. Improvements in experimental techniques lead to a steady acceleration in the gathering of facts about the components of living matter, culminating in Watson and Crick's discovery of the DNA double helix half a century ago, which ushered in the modern era of molecular biology.

The mission of biology remained unchanged during these developments, but knowledge about life became steadily more detailed. As Sommerhoff has remarked, "To put it naïvely, the fundamental problem of theoretical biology is to discover how the behaviour of myriads of blind, stupid, and by inclination chaotic, atoms can obey the laws of physics and chemistry, and at the same time become integrated into organic wholes and into activities of such purpose-like character". Since he wrote those words, experimental molecular biology has advanced far and fast, yet the most important question of all, "what is life?" remains a riddle.

It is a curious fact that although "information" figures so prominently in the central dogma, the concept of information has continued to receive rather cursory treatment in molecular biology textbooks. Even today, the word "information" may not even appear in the index. On the other hand, whole chapters are devoted to energy and energetics, which, like information, is another fundamental, irreducible concept. Although the doctoral thesis of Shannon, one of the fathers of information theory, was entitled "An algebra for theoretical genetics", apart from genetics, biology remained largely untouched by developments in information science.

© Springer-Verlag London 2015
J. Ramsden, *Bioinformatics*, Computational Biology 21,
DOI 10.1007/978-1-4471-6702-0_1

One might speculate on why information was placed so firmly at the core of molecular biology by one of its pioneers. During the preceding decade, there had been tremendous advances in the theory of communication—the science of the transmission of information. Shannon published his seminal paper on the mathematical theory of communication only a few years before Watson and Crick's work. In that context, the notion of a sequence of DNA bases as message with meaning seemed only natural, and the next major development—the establishment of the genetic code with which the DNA sequence could be transformed into a protein sequence—was cast very much in the language and concepts of communication theory. More puzzling is that there was not subsequently a more vigorous interchange between the two disciplines. Probably the lack of extensive datasets and of powerful computers, which made the necessary calculations intolerably tedious, or simply too long, provides sufficient explanation for this neglect—and hence, now that both these requirements (datasets and powerful computers) are being met, it is not surprising that there is a great revival in the application of information ideas to biology. One may indeed hope that this revival will at last lead to a real answer being advanced in response to the vital question "what is life?": In other words, information science is perhaps the missing discipline that, along with the physics and chemistry already being brought to bear, is needed to answer the question.

1.1   What is Bioinformatics?

The term "bioinformatics" seems to have been first used in the mid-1980s in order to describe the application of information science and technology in the life sciences. The definition was at that time very general, covering everything from robotics to artificial intelligence. Later, bioinformatics came to be somewhat prosaically defined as "the use of computers to retrieve, process, analyse, and simulate biological information". An even narrower definition was "the application of information technology to the management of biological data". Such definitions fail to capture the centrality of information in biology. If, indeed, information is the most fundamental concept underlying biology and bioinformatics is the exploration of all the ramifications and implications of that basis, then bioinformatics is excellently positioned to revive consideration of the central question "what is life?" A more appropriate definition of bioinformatics is, therefore, "the science of how information is generated, transmitted, received, stored, processed and interpreted in biological systems" or, more succinctly, "the application of information science to biology".

The emergence of information theory by the middle of the twentieth century enabled the creation of a formal framework within which information could be quantified. To be sure, the theory was, and to some extent still is, incomplete, especially regarding those aspects going beyond the merely faithful transmission of messages, in order to enquire about, and even quantify, the meaning and significance of messages.

In parallel to these developments, other advances, including the development of the idea of algorithmic complexity, with which the names of Kolmogorov and Chaitin are associated, allowed a number of other crucial clarifications to be made, including the notion that randomness is minimally informative. The DNA sequence of a living organism must depart in some way from randomness, and the study of these departures could be said to constitute the core of bioinformatics.

Alongside information theory, cybernetics developed as a distinctive science at around the same time and largely within the same constellation. Its definition is well conveyed by the subtitle of Wiener's eponymous book (1948): "the study of control and communication in the animal and the machine". The word itself was coined by Ampère (as *cybernétique*) more than a century earlier. It is derived from the Greek $\kappa\upsilon\beta\epsilon\rho\nu\eta\tau\zeta\sigma$, meaning steersman, from which we get our Latin *gubernetes*, morphing into "governor". A governor such as Watts' for the steam engine uses a relatively simple feedback mechanism in its operation, and feedback has remained an important concept within cybernetics. It appears to have already been used by Plato as a metaphor for governance in society (which was the interest of Ampère in the topic). According to Aristotle, $\kappa\upsilon\beta\epsilon\rho\nu\eta\tau\iota\kappa\eta\ \tau\eta\chi\nu\epsilon$, the art of the steersman, implied teleological (goal-oriented) activity as well as knowledge, which is, as Sommerhoff has pointed out, perhaps the most characteristic apparent feature of living organisms. Information is, of course, central to considering how control and communication are enacted and, hence, bioinformatics and cybernetics become almost synonymous.

1.2   What Can Bioinformatics Do?

In a very short interval, "bioinformatics" has become an extremely active research field. Although it began with sequence comparison (which is a subbranch of the study of the nonrandomness of DNA sequences), it now encompasses a far wider spread of activity, which truly epitomizes modern scientific research. It is highly interdisciplinary, requiring at least mathematical, biological, physical, and chemical knowledge, and its implementation may furthermore require knowledge of computer science, chemical engineering, biotechnology, medicine, pharmacology, etc. There is, moreover, little distinction between work carried out in the public domain, either in academic institutions (universities) or state research laboratories, or privately by commercial firms.

The handling and analysis of DNA sequences remains one of the prime tasks of bioinformatics. This topic is usually divided into two parts: (1) functional genomics, which seeks to determine the rôle of the sequence in the living cell, either as a transcribed and translated unit (i.e., a protein, the description of the function of which might involve knowledge of its structure and potential interactions) or as a regulatory motif, whether as a promoter site or as a short sequence transcribed as a piece of small interfering RNA; and (2) comparative genomics, in which the sequences from different organisms, or even different individuals, are compared in order to determine ancestries and correlations with disease. Clearly, the comparison of unknown sequences with known ones can also help to elucidate function; both parts are concerned with the search for patterns or *regularities*—which is indeed the core of all scientific work. One can feel that it is fortunate (for scientists) that life is in some sense encapsulated in such a highly formalized object as a sequence of symbols (a string).

The requirement of entire genomes to feed this search has led to tremendous advances in the technology of rapid sequencing, which, in turn, has put new demands on informatics for interpreting the raw output of a sequencer. If a DNA sequence is the message, then functional genomics is concerned with the meaning of the message and, in turn, this has led to the experimental analysis of the RNA transcripts (the transcriptome) and the repertoire of expressed proteins (the proteome), each of which presents fresh informatics challenges. They have themselves spawned interest in the products of protein activity—saccharides (glycomics), lipids (lipidomics), and metabolites (metabolomics). All these "-omics", including the integrative phenomics, are considered to be part of bioinformatics and are covered in this book. Mindful of the need to keep the length of this book within reasonable bounds, chemical genomics (or chemogenomics), defined as the use of small molecules to study the functions of the cell at the genome level (including investigation of the effects of such molecules on gene expression), although closely related to the other topics, is not covered. Computational biology (defined as the application of quantitative and analytical techniques to model biological systems), is only covered via a brief consideration of the virtual living organism. Also in order to keep the length of this book within reasonable bounds, the impressive attempts of Holland, Ray and others to model some characteristic features of life—speciation and evolution—entirely *in silico* using digital organisms (i.e., computer programs able to self-replicate, mutate, etc.) are not covered.

Many bioinformaticians wonder what is the relation of their field to systems biology, which "aims to understand biological behaviour at the systems level through an abstract description in terms of mathematical and computational formalisms".[1] As far as can be discerned ("definitions" abound), it is really a subset of bioinformatics dealing especially with modelling and perhaps constituting the intersection of bioinformatics with computational biology. If emphasis is placed on the abstract description aspect, systems biology would appear to be the same as what was previously called analytical biology.

Aside from sequencing, another product of high-throughput biology is the experimental determination of interactions between objects (i.e., between genes, proteins and metabolites)—now called interactomics—and the inference of regulatory networks from such data has also become a significant part of bioinformatics.

It seems perfectly reasonable to include neurophysiology within bioinformatics, since it deals with how information is generated, transmitted, received, and interpreted in the brain; that is, it corresponds precisely with our definition given above, although it is often considered to be a vast field in its own right. This is even more

---

[1] Kolch et al. (2005).

true of the science of human communication and cognition, which has, regrettably to be left aside in this book.

The book is organized into three main parts. Part I deals, largely heuristically, with the concept of information and some essential basic knowledge associated with it—what one needs to know in order to make sense of the application of information theory to biology—including elements of combinatorics and probability theory, and of pattern recognition and clustering. It would have been quite appropriate to have included a chapter on statistical models since they are needed in much work dealing with large quantities of biological data, and only the unreasonable expansion of the book that such inclusion would have implied, and the availability of good books on the topic, prevented it. Part II is a compact primer on biology, both molecular and organismal. It includes formal aspects of mechanism, whether living or not, such as regulation and adaptation. Part III deals with applications; that is, areas of active current work, including genomics and toxicogenomics, proteomics and interactomics (the study of the repertoire of molecular interactions in a cell). Topics such as practical programming, or database handling, are left out since there are already several excellent books available covering them.[2] A similar remark applies to such topics as the design of genetic association studies.

Although the gene has been at the heart of bioinformatics from the beginning, the main challenge seems now to lie in understanding the functional relationships between biological objects beyond those encoded in the nucleotide sequence. This zone is called epigenetics, and we are only just entering it. It still appears mostly formless, with tantalizing, but ever more frequent, glimpses of incredible complexity, and if there are clues to its structure in the nucleotide sequence, they remain as yet largely hidden from us.

Attention should be called to the fact that for various reasons, including experimental ones, the usual procedure in the physical sciences, which is first to assign numbers to the phenomenon under investigation and then to manipulate the numbers according to the usual rules of mathematics, both operations being publicly declared and publicly accessible, is often confounded in the biological sciences, not least because of the great complexity of the phenomena under investigation. Bioinformatics may be able to provide the needed quantification for the vast tracts of biology where it is so sorely needed.

One consequence of the apparent reluctance of experimenters in the biological sciences to assign numbers to the phenomena they investigate is that the experimental literature is very wordy and hence voluminous, so much so that a subbranch of bioinformatics called text mining has grown up, whose aim is to automatically extract information from published articles, from which, for example, the association of a pair of genes can be inferred. The techniques involved are essentially the same as those involved in searching for genes in a DNA sequence. They are briefly discussed in the final chapter.

---

[2]The development of new algorithms and statistics is, of course, in itself an important branch of bioinformatics.

Activity in a new field begins with the advanced researcher, later it becomes material suitable for doctoral theses, and finally becomes part of undergraduate studies. Bioinformatics seems to be on the threshold of the shift into undergraduate work. The enormous virgin fields opened up by the sequencing of the entire DNA of organisms has imparted tremendous impetus and urgency, and practitioners are now required at every level, from the implementation of the latest findings in medicine and ecology to the continued pushing back of the frontiers of knowledge.

References

Kolch W, Calder M, Gilbert D (2005) When kinases meet mathematics. FEBS Lett 579:1891–1895

Wiener N (1948) Cybernetics, or control and communication in the animal and the machine. Actualités Sci. Ind. no 1053. Hermann & Cie, Paris

Part I
Information

The Nature of Information

2

What is information? We have already asserted that it is a profound, primitive (i.e., irreducible) concept. Dictionary definitions include "(desired) items of knowledge"; for example, one wishes to know the length of a piece of wood. It appears to be less than a foot long, so we measure it with our desktop ruler marked off in inches, with the result, let us say, "between six and seven inches". This result is clearly an item of desired knowledge, hence information. We shall return to this example later. Another definition is "fact(s) learned about something", implying that there is a definable object to which the facts are related, suggesting the need for context and meaning. A further definition is "what is conveyed or represented by a particular arrangement of things"; the dots on the head of a matrix printer shape a letter, the bar code on an item of merchandise represents facts about the nature, origin, and price of the merchandise, and a sequence of letters can convey a possibly infinite range of meanings. A thesaurus gives as synonyms "advice, data, instruction, message, news, report". Finally, we have "a mathematical quantity expressing the probability of occurrence of a specific sequence of symbols or impulses as against that of other sequences (i.e., messages)". This definition links the quantification of information to a probability, which, as we shall see, plays a major rôle in the development of the subject.

We also note that "information science" is defined as the "study of processes for storing and retrieving information", and "information theory" is defined as the "quantitative study of transmission processes for storing and retrieving of information by signals"; that is, it deals with the mathematical problems arising in connexion with the storage, transformation, and transmission of information. This forms the material for Chap. 3. Etymologically, the word "information" comes from the Latin *forma*, form, from *formare*, to give shape to, to describe.

Most information can be reduced to the response, or series of responses, to a question, or series of questions, admitting only yes or no as an answer. We call these yes/no, or dichotomous, questions. Typically, interpretation depends heavily on

© Springer-Verlag London 2015
J. Ramsden, *Bioinformatics*, Computational Biology 21,
DOI 10.1007/978-1-4471-6702-0_2

context.[1] Consider a would-be passenger racing up to a railway station. His question "has the train gone?" may indeed be answered by "yes" or "no"—although, in practice, a third alternative, "don't know," may be encountered. At a small wayside station, with the traveller arriving within five minutes of the expected departure time of the only train scheduled within the next hour, the answer (yes or no) would be unambiguous and will convey exactly one bit of information, as will be explained below. If we insist on the qualification "desired," an unsolicited remark of the stationmaster, "the train has gone," may or may not convey information to the hopeful passenger. Should the traveller have seen with his own eyes the train depart a minute before, the stationmaster's remark would certainly not convey any information.

Consider now a junction at which, after leaving the station, the lines diverge in three different directions. The remark "the train has gone", assuming the information was desired, would still convey one bit of information, but by in addition specifying the direction, viz. "the train has gone to X," or "the train to X has gone," "X" being one of the three possible destinations, the remark would convey $\log_2 3 = 1.59$ bits of information, this being the average number of questions admitting yes/no answers required to specify the fact of departure to X, as opposed to either of the two other directions.

This little scenario illustrates several crucial points:

1. Variety exists. In a formless, amorphous world there is no information to convey.
2. The amount of information received depends on what the recipient knows already.
3. The amount of information can only be calculated if the set of possible messages (responses) has been predefined.

Dichotomous information often has a hierarchical structure; for example, on a journey, a selection of direction has to be made at every cross-road. Given an ultimate destination, successive choices are only meaningful on the basis of preceding ones. Consider also an infant, who "chooses" (according to its environment) which language it will speak. As an adolescent, he chooses a profession, again with an influence from the environment and, in making this choice, knowledge of a certain language may be primordial. As an adult there will be further career choices, which will usually be intimately related to the previous choice of a profession.

Let us now reexamine the measurement of the length of a stick. It must be specified in advance that it does not exceed a certain value—say one foot. This will suffice to allow an appropriate measuring tool to be selected. If all we had was a measuring stick exactly one foot long, we could simply ascertain whether the unknown piece was longer or shorter, and this information would provide one bit of information, if any length was *a priori* possible for the unknown piece.

Suppose, however, that the measuring stick is marked off in 1-inch divisions. If the probabilities p of the unknown piece being any particular length l (measured to

---

[1]When it comes to the quantification of information, context is usually formalized through the provision of a finite set of possible answers (choices). See Sect. 2.3.2.

the nearest inch), with $0 < l \leq 12$, were *a priori* equal (i.e., $p = \frac{1}{12}$ for each possible length), then the information produced by the measurement equals $\log_2 12 = 3.59$ bits, this being the average number of questions admitting yes/no answers required to specify the length to the nearest inch, as the reader may verify. On the other hand, were we to have some prior information, according to which we had good reason to suppose the length to be close to 9 inches (perhaps we had previously requested the wood to be chopped to that length), the probabilities of the lengths 8, 9, and 10 inches would perhaps be 0.25 each, and the sum of all the others would be 0.25. The existence of this prior knowledge would somewhat reduce the quantity of information gained from the measurement, namely to $\frac{3}{4} \log_2 4 + \frac{1}{4} \log_2 36 = 2.79$ bits. Should the ruler have been marked off in tenths of an inch, the measurement would have yielded considerably more information, namely $\log_2 120 = 6.91$ bits, assuming all the probabilities of the wood being any particular length to be equal (i.e., $\frac{1}{120}$ each).

Variety One of the most striking characteristics of the natural, especially the living, world around us is its variety. This variety stands in great contrast to the world studied by the methods of physics and chemistry, in which every electron and every proton (etc.) in the universe are presumed to be identical, and we have no evidence to gainsay this presumption. Similarly, every atom of helium (^{4}He) is similar to every other one, and indeed it is often emphasized that chemistry could only make progress as a quantitative science after the realization that pure substances were necessary for the investigation of reactions and the like, such that a sample of naphthalene in a laboratory in Germany would behave in precisely the same way as one in Japan.

If we are shown a tray containing balls of three colours, red (r), blue (b), and white (w), we might reasonably assert that the variety is three. Hence, one way to quantify variety is simply to count the number of different kinds of objects. Thus, the variety of either of the sets {r, b, w} and {r, b, b, r, w, r, w, w, b} is equal to three; the set {r, r, w, w, w} has a variety of only two, and so forth. The objects considered should of course be in the same category; that is, if the category were specified as "ball," then we would have difficulty if the tray also included a banana and an ashtray. However, one could then redefine the category.

If there were only one kind of ball, say red, then our counting procedure would yield a variety of one. It is more natural, however, to say that there is no variety if all the objects are the same, suggesting that the logarithm of the number of objects is a more reasonable way to quantify variety. If all the objects are the same, the variety is then zero. We are, of course, at liberty to choose any base for the logarithm; if the base is 2, then conventionally the variety is given in units of bits, a contraction of *bi*nary *dig*it. Hence, two kinds of objects have a variety of $\log_2 2 = 1$ bit, and three kinds give $\log_2 3 = \frac{\log_{10} 3}{\log_{10} 2} = \frac{0.477}{0.301} = 1.58$ bits. The variety in bits is the average number of yes/no questions required to ascertain the number of different kinds of objects or to identify the kind of any object chosen from the set.[2]

---

[2]This primitive notion of variety is related to the diversity measured by biometricians concerned with assessing the variety of species in an ecosystem (biocoenosis). Diversity D is essentially variety

The Shannon Index The formula that we used to determine the quantity I of information delivered by a measurement that fixes the result as one out of n equally likely possibilities, each having a probability $p_i, i = 1, \ldots, n$, all equal to $1/n$, was

$$I = -\log p = \log n .\tag{2.4}$$

It is called Hartley's formula. If the base of the logarithm is 2, then the formula yields numerical values in bits. Where the probabilities of the different alternatives are not equal, then a weighted mean must be taken:

$$I = -\sum_{i=1}^{n} p_i \log_2 p_i .\tag{2.5}$$

This generalization is called the Shannon or Shannon–Wiener index. In other words, the quantity of information is weighted logarithmic variety. Note that the quantity of information given by Eq. (2.5) is always less than that given by the equiprobable case (2.4). This follows from Jensen's inequality.[3]

Why is the negative of the sum taken? I in fact represents the *gain* of information due to the measurement. In general,

$$\text{gain (in something)} = \text{final value} - \text{initial value} .\tag{2.7}$$

The initial value represents the uncertainty in the outcome *prior* to the measurement. Shannon (1951) takes the *final* value (i.e., the result of the measurement), to be a single value with variety one, hence using (2.5), $I = 0$ after the measurement; that is, he considers the result to be known with certainty once it has been delivered.

(Footnote 2 continued)
weighted according to the relative abundances (i.e., probability p_i of occurrence) of the N different types, and this can be done in different ways. Parameters in use by practitioners include

$$D_0 = N \qquad \text{(no weighting)},\tag{2.1}$$

$$D_1 = \exp(I) \qquad \text{(the exponential of Shannon's index)},\tag{2.2}$$

$$D_2 = 1/\sum_{i=1}^{N} p_i^2 \quad \text{(the reciprocal of Simpson's index)}.\tag{2.3}$$

[3]If $g(x)$ is a convex function on an interval (a, b), if $x_1, x_2, \ldots, x_n$ are arbitrary real numbers $a < x_k < b$, and if $w_1, w_2, \ldots, w_n$ are positive numbers with $\sum_{k=1}^{n} w_k = 1$, then

$$g\left(\sum_{k=1}^{n} w_k x_k\right) \leq \sum_{k=1}^{n} w_k g(x_k) .\tag{2.6}$$

Inequality (2.6) is then applied to the convex function $y = x \log x$ $(x > 0)$ with $x_k = p_k$ and $w_k = 1/n$ $(k = 1, 2, \ldots, n)$ to get $I(p_1, p_2, \ldots, p_n) \leq \log n$.

Hence, it is considered to have zero information, and it is in this sense that an information processor is also an information annihilator. Wiener (1948) considers the more general case in which the result of the measurement could be less than certain (e.g., still a distribution, but narrower than the one measured).

The gain of information I is equivalent to the removal of uncertainty; hence, information could be defined as "that which removes uncertainty". It corresponds to the reduction of variety perceived by an observer and is inversely proportional to the probability of a particular value being read, or a particular symbol (or set of symbols) being selected, or, more generally, is inversely proportional to the probability of a message being received and remembered.

Example. An $N \times N$ grid of pixels, each of which can be either black or white, can convey at most $-\sum_i^{N^2} \frac{1}{2} \log_2 \frac{1}{2}$ bits of information. This maximum is achieved when the probability of being either black or white is equal.

I defined by Eqs. (2.4) and (2.5) has the properties that one may reasonably postulate should be possessed by a measure of information, namely

1. $I(E_{NM}) = I(E_N) + I(E_M)$, for $N, M = 1, 2, \ldots$;
2. $I(E_N) \leq I(E_{N+1})$;
3. $I(E_2) = 1$.

Example. How much information is contained in a sequence of DNA? If each of the four bases are chosen with equal probability (i.e., $p = \frac{1}{4}$), the information in a decamer is $10 \log_2 4 = 20$ bits. It is the average number of yes/no questions that would be needed to ascertain the sequence. If the sequence were completely unknown before questioning, this is the gain in information. Any constraints imposed on the assembly of the sequence—for example, a rule that "AA" is never followed by "T", will lower the information content of the sequence (i.e., the gain in information upon receiving the sequence, assuming that those constraints are known to us). Some proteins are heavily constrained; the antifreeze glycoprotein (alanine-alanine-threonine)$_n$ could be simply specified by the instruction "repeat AAT n times", much more compactly than writing out the amino acid sequence in full, and the quantity of information gained upon being informed of the sequence is correspondingly small.

Thermodynamic Entropy One often encounters the word "entropy" used synonymously with information (or its removal). Entropy (S) in a physical system represents the ability of a system to absorb energy without increasing its temperature. Under isothermal conditions (i.e., at a constant temperature T),

$$dQ = T \, dS \,, \tag{2.8}$$

where dQ is the heat that flows into the system. In thermodynamics, the internal energy E of a system is formally defined by the First Law as the difference between the heat and dW, the work done by the system:

$$dE = dQ - dW \,. \tag{2.9}$$

The only way that a system can absorb heat without raising its temperature is by becoming more disordered. Hence, entropy is a measure of disorder. Starting from a microscopic viewpoint, entropy is given by the famous formula inscribed on Boltzmann's tombstone:

$$S = k_B \ln W \, , \tag{2.10}$$

where k_B is his constant and W is the number of (micro)states available to the system. Note that reducing the number of states reduces the disorder. Information amounting to $\log_2 W$ bits is required to specify one particular microstate, assuming that all microstates have the same probability of being occupied, according to Hartley's formula; the specification of a particular microstate removes that amount of uncertainty. Thermodynamical entropy defined by Eq. (2.8), statistical mechanical entropy (2.10), and the Hartley or Shannon index only differ from each other by numerical constants.

Although the set of positions and momenta of the molecules in a gas at a given instant can thus be considered as information, within a microscopic interval (between atomic collisions, of the order of 0.1 ps) this set is forgotten and another set is realized. The positions and momenta constitute microscopic information; the quantity of macroscopic (remembered) information is zero. In general, the quantity of macroinformation is far less than the quantity of (forgotten) microinformation, but the former is far more valuable.[4]

In the world of engineering, this state of affairs has of course always been recognized. One does not need to know the temperature (within reason!) in order to design a bridge or a mechanism. The essential features of any construction are found in a few large-scale correlated motions; the vast number of uncorrelated, thermal degrees of freedom are generally unimportant.

Symbol and Word Entropies The Shannon index (2.5) gives the average information per symbol; an analogous quantity I_n can be defined for the probability of n-mers (n-symbol "words"), whence the differential entropy $\tilde{I}_n$,

$$\tilde{I}_n = I_{n+1} - I_n \, , \tag{2.11}$$

whose asymptotic limit ($n \to \infty$) Shannon calls "entropy of the source", is a measure of the information in the $(n+1)$th symbol, assuming the n previous ones are known. The decay of $\tilde{I}_n$ quantifies correlations within the symbolic sequence (an aspect of memory).

---

[4]"Forgetting" implies decay of information; what does "remembering" mean? It means to bring a system to a defined stable state (i.e., one of two or more states), and the system can only switch to another state under the influence of an external impulse. The physical realization of such systems implies a minimum of several atoms; as a rule a single atom, or a simple small molecule, can exist in only one stable state. Among the smallest molecules fulfilling this condition are sugars and amino acids, which can exist in left- and right-handed chiralities. Note that many biological macromolecules and supramolecular assemblies can exist in several stable states.

2.1   Structure and Quantity

In our discussion so far we have tacitly assumed that we know *a priori* the set from which the actual measurement will come. In an actual physical experiment, this is like knowing from which dial we shall take readings of the position of the pointer, for example, and, furthermore, this knowledge may comprise all the information required to construct and use the meter, which is far more than that needed to formally specify the circuit diagram and other details of the construction. It would also have to include blueprints for the machinery needed to make the mechanical and electronic components, for manufacturing the required materials from available matter, and so forth. In many cases we do not need to concern ourselves about all this, because we are only interested in the gain in information (i.e., loss of uncertainty) obtained by receiving the result of the dial reading, which is given by Eq. (2.5). The information pertinent to the construction of the experiment usually remains the same, hence cancels out (Eq. 2.7). In other words, the Shannon–Weaver index is strictly concerned with the metrical aspects of information, not with its structure.

2.1.1   The Generation of Information

Prior to carrying out an experiment, or an observation, there is objective uncertainty due to the fact that several possibilities (for the result) have to be taken into account. The information furnished by the outcome of the experiment reduces this uncertainty: R.A. Fisher (1951) defined the quantity of information furnished by a series of repeated measurements as the reciprocal of the variance:

$$I_F(x) \leq 1/\langle(x_{est} - x)^2\rangle \tag{2.12}$$

where I_F is the Fisher information and the denominator of the right-hand side is the variance of the estimator x_{est}.[5] One use of I_F is to measure the encoding accuracy of a population of neurons subject to some stimulus (Chap. 17); maximizing I_F optimizes extraction of the value of the stimulus.[6]

2.1.2   Conditional and Unconditional Information

Information about real events that have happened (e.g., a volcanic eruption), or about entities that exist (e.g., a sequence of DNA) is primarily unconditional; that is, it does not depend on anything (as soon as information is encoded, however, it becomes conditional on the code).

---

[5]The relation between the Shannon index and Fisher's information, which refers to the intrinsic accuracy of an experimental result, is treated by Kullback and Leibler (1951).
[6]An example is given by Karbowski (2000).

Scientific work has two stages:

1. Receiving unconditional information from nature (by making observations in the field, doing experiments in the laboratory).
2. Generating conditional information in the form of hypotheses and theories relating the observed facts to each other using axiom systems. The success of any theory (which may be one of several) largely depends on general acceptance of the chosen propositions and the mathematical apparatus used to manipulate the elements of the theory; that is, there is a strongly social aspect involved.

Conditional information tends to be unified; for example, a group of scattered tribes, or practitioners of initially disparate disciplines, may end up speaking a common language (they may then comprehend the information they exchange as being unconditional and may ultimately end up believing that there cannot be other languages). Encoded information is conditional on agreement between emitters and receivers concerning the code.

2.1.3 Experiments and Observations

Consider once again the example of the measurement of the length of an object using a ruler and the information gained thereby. The gain presupposes the existence of a world of objects and knowledge, including the ruler itself and its calibration in appropriate units of measurement. The overall procedure is captured, albeit imperfectly, in Fig. 2.1.

The essential point is that "information" has two parts: a prior part embodied by the physical apparatus, the knowledge required to carry out the experiment or observation, and so forth; and a posterior part equal to the loss in uncertainty about the system due to having made the observation. The prior part can be thought of as specifying the set of possible values from which the observed value must come. In a physical measurement, it is related to the structure of the experiment and the instruments it employs, and the millennia of civilization that have enabled such activities. The posterior part (I) is sometimes called "missing information" because once the prior part (K) is specified, the system still has the freedom, quantified by I, to adopt different microstates. In a musical analogy, K would correspond to the structure of a Bach fugue and I to the freedom the performer has in making

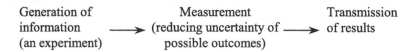

Generation of   Measurement   Transmission
information   ⟶   (reducing uncertainty of ⟶   of results
(an experiment)   possible outcomes)

Fig. 2.1 The procedures involved in carrying out an experiment, from conception to ultimate dissemination

interpretational choices while still respecting the structure.[7] One could say that the magnitude of I corresponds to the degree of logical indeterminacy inhering in the system, in other words that part of its description that cannot be formulated within itself; it is the amount of *selective* information lacking.

I can often be calculated according to the procedures described in the previous section (the Hartley or Shannon index). If we need to quantify K, it can be done using the concept of algorithmic information content (AIC) or Kolmogorov information, which corresponds to the length of the most concise description of what is known about the system (see Sect. 6.5). Hence, the total information $\mathfrak{I}$[8] is the sum of the ensemble (Shannon) entropy I and the physical (Kolmogorov) entropy K:

$$\mathfrak{I} = I + K . \tag{2.13}$$

Mackay (1950) proposed the terms "logon" for the structural (prior) information, equivalent to K in Eq. (2.13), and "metron" for the metrical (posterior) measurement. The gain in information from a measurement (Eq. 2.7) falls wholly within the metrical domain, of course, and within that domain, there is a prior and posterior component (cf. Sect. 5.4).

To summarize, the Kolmogorov information K can be used to define the structure of information and is calculated by considering the system used to make a measurement. The result of the measurement is macroscopic, remembered information, quantified by the Shannon index I. The gain in information equals ($\text{final}_f - \text{initial}_i$ information):

$$I = (I_f + K) - (I_i + K) = I_f - I_i . \tag{2.14}$$

In other words, it is unexceptionable to assume that the measurement procedure does not change the structural information, although this must only be regarded as a cautious, provisional statement. Presumably, any measurement or series of measurements that overthrows the theoretical framework within which a measurement was made does actually lead to a change in K. Equation (2.13) formalizes the notion of quiddity *qua* essence, comprising substance (K) and properties (I). The calculation of K will be dealt with in more detail in Chap. 6. As a final remark in this section, we note that the results of an experiment or observation transmitted elsewhere may have the same effect on the recipient as if he had carried out the experiment himself.

Problem. Critically scrutinize Fig. 2.1 in the light of the above discussion and attempt to quantify the information flows.

2.2   Constraint

Shannon puts emphasis on the information resulting from the selection from a set of possible alternatives (implying the existence of alternatives)—information can only be received where there is doubt. Much of the theory of information deals with

---

[7]Cf. Tureck (1995).
[8]Called the physical information of a system by Zurek (1989).

signals, which operate on the set of alternatives constituting the recipient's doubt to yield a lesser doubt, or even certainty (zero doubt). Thus, the signals themselves have an information content by virtue of their potential for making selections; the quantity of information corresponds to the intensity of selection or to the recipient's surprise upon receiving the information. I from Eq. (2.5) gives the average information content per symbol; it is a weighted mean of the degree of uncertainty (i.e., freedom of choice) in choosing a symbol before any choice is made.

If we are writing a piece of prose, and even more so if it is verse, our freedom of choice of letters is considerably constrained; for example, the probability that "x" follows "g" in an English text is much lower than $\frac{1}{26}$ (or $\frac{1}{27}$ if we include, as we should, the space as a symbol). In other words, the selection of a particular letter depends on the preceding symbol, or group of preceding symbols. This problem in linguistics was first investigated by Markov, who encoded a poem of Pushkin's using a binary coding scheme admitting consonants (C) or vowels (V). Markov (1913) proposed that the selection of successive symbols C or V no longer depended on their probabilities as determined by their frequencies ($v = V/(V + C)$, where V and C are, respectively, the total numbers of vowels and consonants). To every pair of letters (L_j, L_k) there corresponds a conditional probability p_{jk}; given that L_j has occurred, the probability of L_k at the next selection is p_{jk}. If the initial letter has a probability a_j, then the probability of the sequence $(L_j, L_k, L_l) = a_j p_{jk} p_{kl}$ and so forth. The scheme can be conveniently written in matrix notation:

$$
\begin{array}{c|cc}
\rightarrow & C & V \\
\hline
C & p_{cc} & p_{cv} \\
V & p_{vc} & p_{vv}
\end{array}
\tag{2.15}
$$

where p_{cc} means the probability that a consonant is followed by another consonant, and similarly for the other terms. The matrix is stochastic; that is, the rows must add up to 1. If every column is identical, then there is no dependence on the preceding symbol, and we revert to a random, or zeroth-order Markov, process. Suppose now that observation reveals that the probability of C occurring after V preceded by C is different from that of C occurring after V preceded by V, or even that the probability of C occurring after VV preceded by C is different from that of C occurring after VV preceded by V. These higher-order Markov processes can be recoded in strict Markov form; thus, for the second-order process (dependency of the probabilities on the two preceding symbols) "VVC" can be written as a transition from VV to VC, and hence the matrix of transition probabilities becomes

$$
\begin{array}{c|cccc}
\rightarrow & CC & CV & VC & VV \\
\hline
CC & p_{ccc} & p_{ccv} & 0 & 0 \\
CV & 0 & 0 & p_{cvc} & p_{cvv} \\
VC & p_{vcc} & p_{vcv} & 0 & 0 \\
VV & 0 & 0 & p_{vvc} & p_{vvv}
\end{array}
\tag{2.16}
$$

and so on for higher orders. Notice that some transitions necessarily have zero probability.[9]

The reader may object that one rarely composes text letter by letter, but rather word by word. Clearly, there are strong constraints governing the succession of words in a text. The frequencies of these successions can be obtained by counting word occurrences in very long text and are then used to construct the transition matrix, which is, of course, gigantic even for a first-order process. We remark that a book ending with "... in the solid state is greatly aided by this new tool" is more likely to begin with "Rocket motor design received a considerable boost when ..." than one ending "I became submerged in my thoughts which sparkled with a cold light".[10]

We note here that clearly one may attempt to model DNA or protein sequences as Markov processes, as will be discussed in Part III. Markov chains as such will be discussed more fully in Chap. 6.

The notion of constraint applies whenever a set "is smaller than it might be". The classic example is that of road traffic lights, which display various combinations of red, amber, and green, each of which may be on or off. Although $2^3 = 8$ combinations are theoretically possible, in most countries only certain combinations are used, typically only four out of the eight. Constraints are ubiquitous in the universe and much of science consists in determining them; thus, in a sense, "constraint" is synonymous with "regularity". Laws of nature are clearly constraints, and the very existence of physical objects such as tables and aeroplanes, which have fewer degrees of freedom than their constituent parts considered separately, is a manifestation of constraint.

In this book we are particularly concerned with constraints applied to sequences. Clearly, if a Markov process is in operation, the variety of the set of possible sequences generated from a particular alphabet is smaller than it would be had successive symbols been freely selected; that is, it is indeed "smaller than it might have been". "Might have been" requires the qualification, then, of "would have been if successive symbols had been freely (or randomly—leaving the discussion of 'randomness' to Chap. 6) selected". We already know how to calculate the entropy (or information, or Shannon index, or Shannon–Weaver index) I of a random sequence (Eq. 2.5); there is a precise way of calculating the entropy per symbol for a Markov process (see Sect. 6.2), and the reader may use the formula derived there to verify that the entropy of a Markov process is less than that of a "perfectly random" process. Using some of the terminology already introduced, we may expand on this statement to say that the surprise occasioned by receiving a piece of information is lower if constraint is operating; for example, when spelling out a word, it is practically superfluous to say "u" after "q".

The constraints affecting the choice of successive words are a manifestation of the syntax of a language. In the next chapter other ways in which constraint can

---

[9]See also Sect. 6.2.

[10]Good (1969) has shown that ordinary language cannot be represented even by a Markov process of infinite order.

operate will be examined, but for now we can simply state that whenever constraint is present, the entropy (of the set we are considering, hence of the information received by selecting a member of that set) is lower than it would be for a perfectly random selection from that set.

This maximum entropy (which, in physical systems, corresponds to the most probable arrangement; i.e., to the macroscopic state that can be arranged in the largest number of ways)—let us call it I_{max}—allows us to define a relative entropy I_{rel},

$$I_{rel} = \frac{\text{actual entropy}}{I_{max}} , \qquad (2.17)$$

and a redundancy R,

$$R = 1 - I_{rel} . \qquad (2.18)$$

In a fascinating piece of work, Shannon (1951) established the entropy of English essentially through empirical investigations using rooms full of people trying to guess incomplete texts.[11]

More formally, the relative entropy (Kullback–Leibler distance)[12] between two (discrete) distributions with probability functions a_k and b_k is

$$\mathcal{R}(a, b) = \sum_k a_k \log_2(a_k/b_k) . \qquad (2.19)$$

If a_k is an actual distribution of observations, and b_k is a model description approximating to the data,[13] then $\mathcal{R}(a, b)$ is the expected difference (expressed as the number of bits) between encoding samples from a_k using a code based on a and using a code based on b. This can be seen by writing Eq. (2.19) as

$$\mathcal{R}(a, b) = - \sum_k b_k \log_2 a_k + \sum_k a_k \log_2 a_k , \qquad (2.20)$$

where the first term on the right-hand side is called the cross-entropy of a_k and b_k, the expected number of bits required to encode observations from a when using a code based on b rather than a. Conversely, $\mathcal{R}(a, b)$ is the gain in information if a code based on a rather than b is used.

Suppose that $P\{x_1, x_2, \ldots, x_m\}$ is the probability of having a certain pattern (arrangement), or m-gram $x_1, x_2, \ldots, x_m$,[14] assumed to be ergodic (stationary stochastic).[15] Examples could be the English texts studied by Shannon; of particular relevance to the topic of this book is the problem of predicting the nucleic acid base

---

[11] Note that most computer languages lack redundancy—a single wrong character in a program will usually cause the program to halt, or not compile.

[12] Since $\mathcal{R}(a, b) \neq \mathcal{R}(b, a)$, it is not a true metric and is therefore sometimes called "divergence" rather than "distance".

[13] Possibly constructed *a priori*.

[14] See also Sect. 8.2.

[15] See Sect. 6.1.

following a known (sequenced) arrangement. The conditional probability[16] that the pattern $[(m-1)$-gram$]$ $x_1, x_2, \ldots, x_{m-1}$ is followed by the symbol x_m is

$$P\{x_m | x_1, x_2, \ldots, x_{m-1}\} = \frac{P\{x_1, x_2, \ldots, x_{m-1}, x_m\}}{P\{x_1, x_2, \ldots, x_{m-1}\}} . \qquad (2.21)$$

The "m-length approximation" to the entropy S_m, defined as the average uncertainty about the next symbol, is

$$S_m = - \sum_{x_1, x_2, \ldots, x_{m-1}} P\{x_1, x_2, \ldots, x_{m-1}\}$$

$$\times \sum_x P\{x_m | x_1, x_2, \ldots, x_{m-1}\} \log P\{x_m | x_1, x_2, \ldots, x_{m-1}\} . \qquad (2.22)$$

It includes all possible correlations up to length m. Note that the first sum on the right-hand side is taken over all possible preceding sequences, and the second sum is taken over all possible symbols. The *correlation information* is defined as

$$k_m = S_{m-1} - S_m \ (m \geq 2) . \qquad (2.23)$$

S_1 is simply the Shannon information (Eq. 2.5). If the probability of the different symbols is *a priori* equal, then the information is given by Hartley's formula (2.4).[17] For $m = 1$,

$$k_1 = \log n - S_1 \qquad (2.24)$$

is known as the *density information*. By recursion we can then write

$$\mathfrak{I} = S + \sum_{m=1}^{\infty} k_m \qquad (2.25)$$

the total information $\mathfrak{I}$ being equal to $\log n$. The first term on the right gives the random component and is defined as $S = \lim_{m \to \infty} S_m$, and the second one gives the redundancy. For a binary string, $S = 1$ if it is random, and the redundancy equals zero. For a regular string like $\ldots 010101 \ldots$, $S = 0$ and $k_2 = 1$; for a first order Markov chain $k_m = 0$ for all $m > 2$.

2.2.1 The Value of Information

In order to quantify value V, we need to know the goal toward which the information will be used. D.S. Chernavsky (1990) points to two cases that may be considered:

(i) The goal can almost certainly be reached by some means or another. In this case a reasonable quantification is

$$V = \text{(cost or time required to reach goal without the information)}$$

$$- \text{(cost or time required to reach goal with the information)} . \qquad (2.26)$$

---

[16]See Sect. 5.2.2.
[17]The effective measure complexity is the weighted sum of the k_m [viz., $\sum_{m=2}^{\infty}(m-1)k_m$]—see Eq. 6.27.

(ii) The probability of reaching the goal is low. Then it is more reasonable to adopt

$$V = \log_2 \frac{\text{prob. of reaching goal with the information}}{\text{prob. of reaching goal without the information}} \,. \qquad (2.27)$$

With both of these measures, irrelevant information is clearly zero-valued.

Durability of information contributes to its value. Intuitively, we have the idea that the more important the information, the longer it is preserved. In antiquity, accounts of major events such as military victories were preserved in massive stone monuments whose inscriptions can still be read today several thousand years later. Military secrets are printed on paper or photographed using silver halide film and stored in bunkers, rather than committed to magnetic media. We tend to write down things we need to remember for a long time.

The value of information is closely related to the problem of weighing the credibility that one should accord a certain received piece of information. The question of weighting scientific data from a series of measurements was an important driver for the development of probability theory. In 1777, Daniel Bernoulli raised this issue in the context of averaging astronomical data, where it was customary to simply reject data deviating too far from the mean and weight all others equally.[18] Bennett (1988) has proposed that his notion of logical depth (Sect. 6.5) provides a formal measure of value, very much in the spirit of Eqs. (2.26) and (2.27). A sequence of coin tosses formally contains much information that has little value; a table giving the positions of the planets every day for several centuries hence contains no more information than the equations of motion and initial conditions from which it was deduced, but saves anyone consulting it the effort of calculating the positions. This suggests that the value of a message resides not in its information *per se* (i.e., its absolutely unpredictable parts) nor in any obvious redundancy (e.g., repetition), but rather in what Bennett has suggested be called buried redundancy: parts predictable only with considerable effort on the part of the recipient of the message. This effort corresponds to logical depth.

The value of information is also related to the amount already possessed. The same Bernoulli asserted that the value (utility in economic parlance) of an amount m of money received is proportional to $\log[(m + c)/c]$, where c is the amount of money already possessed,[19] and a similar relationship may apply to information.

---

[18]D. Bernoulli, Diiudicatio maxime probabilis plurium observationem discrepantium atque verisimillima inductio inde formanda. Acta Acad. Sci. Imp. Petrop. 1 (1777) 3–23. See also L. Euler, Observationes in praecedentem dissertationem illustris Bernoulli. Acta Acad. Sci. Imp. Petrop. 1 (1777) 24–33.

[19]Cf. Thomas (2010).

2.2.2 The Quality of Information

Quality is an attribute that brings us back to the problem posed by Bernoulli in 1777, namely how to weight observations. If we return to our simple measurement of the length of a piece of wood, the reliability may be affected by the physical condition of the measuring stick, its markings, its origin (e.g., from a kindergarten or from Sèvres), the eyesight of the measurer, and so forth.

2.3 Accuracy, Meaning, and Effect

2.3.1 Accuracy

In the preceding sections, we have focused on the information gained when a certain signal, or sequence of signals, is received. The quantity of this information I has been formalized according to its statistical properties. I is of particular relevance when considering how accurately a certain sequence of symbols can be transmitted. This question will be considered in more detail in Chap. 3. For now, let us merely note that no physical device can discriminate between pieces of information differing by arbitrarily small amounts. In the case of a photographic detector, for example, diminishing the difference will require larger and larger detectors in order to discriminate, but photon noise places an ultimate limitation in the way of achieving arbitrarily small detection.

A communication system depending on setting the position of a pointer on a dial to 1 of 6000 positions and letting the position be observed by the distant recipient of the message through a telescope, while allowing a comfortably large range of signs to be transmitted, would be hopelessly prone to reading errors, and it was long ago realized that far more reliable communication could be achieved by using a small number of unambiguously uninterpretable signs (e.g., signalling flags at sea) that could be combined to generate complex messages.[20]

Practical information space is thus normally discrete; for example, meteorological bulletins do not generally give the actual wind speed in kilometres per hour and the direction in degrees, but refer to 1 of the 13 points of the Beaufort scale and 1 of the 8 compass points. The information space is therefore a finite 2-space with 8×13 elements.

---

[20]The same principle applies, in vastly extended form, to the principal systems of writing extant on Earth. In the Chinese system one character, which may be quite elaborate, represents an entire word, which could itself represent (often in a context-dependent fashion) an entire concept. In the alphabetical system, words are built up from syllables. Where there is no difficulty in perceiving a text in full detail, preferably a whole page at a time, the Chinese system must be superior, having more force of expression and enabling the information to be appraised more rapidly. In other cases, such as transmitting messages long distances through a noisy channel, the alphabetic system has evident advantages.

The rule for determining the distance between two words (i.e., the metric of information space) is most conveniently perceived if the words are encoded in binary form. The Hamming distance is the number of digit places in which the two words differ.[21] This metric satisfies the usual rules for distance; that is, if a, b, and c are three points in the space and $D(a, b)$ is the distance between a and b, then

$$D(a, a) = 0 ;$$
$$D(a, b) = D(b, a) > 0 \quad \text{if } b \neq a ;$$
$$D(a, b) + D(b, c) \geq D(a, c) .$$

Other distances can be defined (see Sect. 13.4.2).

In biology, the question of accuracy refers especially to the replication of DNA, its transcription into RNA, and the translation of RNA into protein. It may also refer to the accuracy with which physiological signals can be transmitted within and between cells.

2.3.2 Meaning

Shannon's theory is not primarily concerned with the question of semantic content (i.e., meaning). In the simple example of measuring the length of a piece of wood, the question of meaning scarcely enters into the discourse. In nearly all of the other cases, where we are concerned with receiving signs, or sequences of symbols, after we have received them accurately we can start to concern ourselves with the question of meaning. The issues can range from simple ones of interpretation to involved and complex ones. An example of the former is the interpretation of the order "Wait!" heard in a workshop. It may indeed mean "pause until further notice", but heard by an apprentice standing by a weighing machine, may well be interpreted as "call out the weight of the object on the weighing pan". An example of the latter is the statement "John Smith is departing for Paris", which has very different connotations according to whether it was made in an airport, a railway station or some other place.

It is easy to show that the meaning contained in a message depends on the set of possible messages. Ashby (1956) has constructed the following example. Suppose a prisoner-of-war is allowed to send a message to his family. In one camp, the message can be chosen from the following set:

I am well
I am quite well
I am not well
I am still alive,

and in another camp, only one message may be sent:

I am well.

---

[21]Cf. J.E. Surrick and L.M. Conant, *Laddergrams*, New York: Sears (1927): "Turn bell into ring in six moves" and so forth; and Sect. 13.4.3.

In both cases, there is implicitly a further alternative—no message at all, which would mean that the prisoner is dying or already dead. In the second camp, if the recipient is aware that only one message is permitted, he or she will know that it encompasses several alternatives, which are explicitly available in the first camp. Therefore, the same message (I am well) can mean different things depending on the set from which it is drawn.

In much human communication, it is the context-dependent difference between explicit and implicit meaning that is decisive in determining the ultimate outcome of the reception of information. In the latter example of the previous paragraph, the context—here provided by the physical environment—endows the statement with a large complement of implicit information, which mostly depends on the mental baggage possessed by the recipient of the information; for example, the meaning of a Chinese poem may only be understandable to someone who has assimilated Chinese history and literature since childhood, and will not as a rule be intelligible to a foreigner armed with a dictionary.

A very similar state of affairs is present in the living cell. A given sequence of DNA will have a well-defined explicit meaning in terms of the sequence of amino acids it encodes, and into which it can be translated. In the eukaryotic cell, however, that amino acid sequence may then be glycosylated and further transformed, but in a bacterium, it may not be; indeed it may even misfold and aggregate—a concrete example of implicit meaning dependent on context.

The importance of context in determining implicit meaning is even more graphically illustrated in the case of the developing multicellular organism, in which the cells are initially all identical; according to chemical signals received from their environment they will develop into different kinds of cells. The meaning of the genotype is the phenotype, and it is implicit rather than explicit meaning, which is, of course, why the DNA sequence of any earthly organism sent to an alien civilization will not allow them to reconstruct the organism. Ultimately, most of the cells in the developing embryo become irreversibly different from each other (differentiation), but while they are still pluripotent, they may be transplanted into regions of different chemical composition and change their fate; for example, a cell from the non-neurogenic region of one embryo transplanted into the neurogenic region of another may become a neuroblast (Sect. 10.8.2). The mechanism of such transformations will be discussed in a little more detail in Chap. 10, but here this type of phenomenon serves to illustrate how the implicit meaning of the genome dominates the explicit meaning. This implicit meaning is called epigenetics,[22] and it seems clear that we will not truly understand life before we have developed a powerful way of treating epigenetic phenomena. Shannon's approach has proved very powerful for treating the problem of the accurate transmission of signals, but at present we do

---

[22]Cf. Sects. 10.8.2 and 10.8.3.

not have a comparable foundation for treating the problem of the precise transfer of meaning.[23]

Even at the molecular level, at which phenotype is more circumscribed and could be considered to be the function (of an enzyme), or simply the structure of a protein, there is presently little understanding of the relation between sequence and function, as illustrated by the thousands of known different sequences encoding the same type of structure and function, or different sequences encoding different structures but the same type of function, or similar structures with different functions.

Part of the difficulty is that the function (i.e., biological meaning) is not so conveniently quantifiable as the information content of the sequence encoding it. Even considering the simpler problem of structure alone, there are various approaches yielding very different answers. Supposing that a certain protein has a unique structure [most nonstructural proteins have, of course, several (at least two) structures in order to function; the best-known example is probably haemoglobin]. This structure could be specified by the coordinates of all the constituent atoms, or the dihedral angles of each amino acid, listed in order of the sequence, and at a given resolution [Dewey (1996, 1997) calls this the algorithmic complexity of a protein; cf. K in Eq. (2.13)]. If, however, protein structures come from a finite number of basic types, it suffices to specify one of these types, which moves the problem back into one dealing with Shannon-type information.

In the case of function, a useful starting point could be to consider the immune system, in which the main criterion of function is the affinity of the antibody (or, more precisely, the affinity of a small region of the antibody) to the target antigen. The discussion of affinity and how affinities can lead to networks of interactions will be dealt with in Chap. 16.

The problem of assigning meaning to a sign, or a message (a collection of signs), is usually referred to as the semantic problem. Semantic information cannot be interpreted solely at the syntactical level. Just as a set of antibodies can be ranked in order of affinity, so may a series of statements be ranked in order of semantic precision; for example, consider the statements:

A train will leave.
A train will leave London today.
An express train will leave London Marylebone for Glasgow at 10:20 a.m. today.

and so on. Postal or e-mail addresses have a similar kind of syntactical hierarchy. Although we are not yet able to assign numerical values to meanings, we can at least order them.

Carnap and Bar-Hillel (1952) have framed a theory, rooted in Carnap's theory of inductive probability, attempting to do for semantics what Shannon did for the technical content of a message. It deals with the semantic content of declarative sentences,

---

[23]Given that translation (from nucleic acid to protein) is involved, the proverb "traduttori traditori" is quite apt.

excluding the pragmatic aspects (dealing with the consequences or value of received information for the recipient). It does not deal with the so-called semantic problem of communication, which is concerned with the identity (or approach thereto) between the intended meaning of the sender and the interpretation of meaning by the receiver: Carnap and Bar-Hillel place this explicit involvement of sender and receiver in the realm of pragmatics.

To gain a flavour of their approach, note that the semantic content of sentence j, conditional on having heard sentence i, is content$(j|i) =$ content$(i \& j) -$ content(i), and their measure of information is defined as information$(i) = -\log_2$ content$(\text{NOT } i)$. They consider semantic noise (resulting in misinterpretation of a message, even though all of its individual elements have been perfectly received) and semantic efficiency, which takes experience into account; for example, a language with the predicates W, M, and C, designating respectively warm, moderate, and cold temperatures, would be efficient in a continental climate (e.g., Switzerland or Hungary) but would become inefficient with a move to the western margin of Europe, since M occurs much more frequently there.

Although the quantification of information is deliberately abstracted from the content of a message, taking content into account may allow much more dramatic compression of a message than is possible using solely the statistical redundancy (Eq. 2.18). Consider how words such as "utilization" may be replaced by "use", appellations such as "guidance counsellor" by "counsellor", and phrases such as "at this moment in time" by "at this moment", or simply "now". Many documents can be thus reduced in length by over two-thirds without any loss in meaning (but a considerable gain in readability). With simply constructed texts, algorithmic procedures for accomplishing this that do not require the text to be interpreted can be devised; for example, all the words in the text can be counted and listed in order of frequency of occurrence, and then each sentence is assigned a score according to the numbers of the highest-ranking words (apart from "and", "that", etc.) it contains. The sentences with the highest scores are preferentially retained.

2.3.3 Effect

A signal may be accurately received and its meaning may be understood by the recipient, but that does not guarantee that it will engender the response desired by the sender. This aspect of information deals with the ultimate result and the possibly far-reaching consequences of a message and how the deduced meaning is related to human purposes. The question of the value of information has already been discussed (Sect. 2.2.1), and operationally it comes close to a quantification of effect.

Mackay has proposed that the quantum of effective information is that amount that enables the recipient to make one alteration to the logical pattern describing his awareness of the relevant situation, and this would appear to provide a good basis for quantifying effect. Suppose that an agent has a state of mind M_1, which comprises certain beliefs, hypotheses, and the like (the prior state). The agent then hears a sentence, which causes a change to state of mind M_2, the posterior state, which stands in readiness to make a response. If the meaning of an item of information is its contribution to the agent's total state of conditional readiness for action and the planning of action (i.e., the agent's conditional repertoire of action), then the effect is the ultimate realization of that conditional readiness in terms of actual action.[24]

As soon as we introduce the notion of a conditional repertoire of action, we see that selection must be considered. Indeed, the three essential attributes of an agent are (and note the parallel with the symbolic level) as follows:

1. A repertoire, from which alternative actions can be selected;
2. An evaluator, which assigns values to different states of affairs according to either given or self-set criteria;
3. A selector, which selects actions increasing a positive evaluation and diminishing deleterious evaluation.

One may compare this procedure with that of evolutionary computation (Sect. 8.1), and, *a fortiori*, with that of evolution itself. Here, the selected actions are used to build up a presence in the repertoire (and, assuming that the repertoire remains constant in size, unselected actions will be diminished).

2.3.4   Significs

As summarized by Welby (1911), significs comprises (a) sense ("in what sense is a word used?"), (b) meaning (the specific sense a word is intended to convey), and (c) significance—the far-reaching consequence, implication, ultimate result, or outcome (e.g., of some event or experience). It therefore includes semantics but goes well beyond it.

Problem. Discuss how the significs of n-grams of DNA and of peptides (regulatory oligopeptides and proteins) could be developed.

2.4   Further Remarks on Information Generation

The exercise of intellect involves both the transformation and generation of information, the latter quite possibly involving the crossing of some kind of logical gap. It is a moot point whether the solution of a set of equations contains more information than the equations, since the solution is implicit (and J.S. Mill insisted that induction, not deduction, is the only road to new knowledge). If it does not, are we then no more complex than a zygote, which apparently contains all the information required to generate a functional adult?

---

[24]Wiener subsumes effect into meaning in his definition of "meaningful information".

The reception of information is equivalent to ordering (i.e., an entropy decrease) and corresponds to the various ordering phenomena seen in nature. Three categories can be distinguished:

1. Order from disorder [sometimes called "self-organization" (see also Sect. 9.8)[25]];
2. Order from order (a process based on templating, such as DNA replication or transcription);
3. Order from noise (microscopic information is given macroscopic expression).

The only meaningful way of interpreting the first category is to suppose that the order was implicit in the initial state; hence, it is questionable whether information has actually been generated. In the second category, the volume of ordering has increased, but inevitably at the expense of more disorder elsewhere, because of the physical exigencies of the copying process.[26] Note that copying *per se* does not lead to an increase in the amount of information. The third category is of genuine interest, for it illuminates problems such as that of the development of the zygote, in which environmental information is given meaningful macroscopic expression, such that we are indeed more complex than the zygotes whence we sprang.

Problem. Examine the proposition that the production and dissemination of copies of a document reporting new facts does not increase the amount of information.

2.5 Summary

Information is that which removes uncertainty. It has two aspects: form (what we already know about the system) and content, the result of an operation (e.g., a measurement) carried out within the framework of our extant knowledge. Form specifies the structure of the information. This includes the specification of the set of possible messages that we can receive or the (design and fabrication of and way of using) the instrument used to measure a parameter of the system. It can be quantified as the length of the shortest algorithm able to specify the system (Kolmogorov information). If we know the set from which the result of the measurement operation has to come, the (metrical) content of the operation is given by the Shannon index (reducing to the Hartley index if the choices are equiprobable). A message (e.g., a succession of symbols) that directs our selection is, upon receipt, essentially equivalent to the result of the measurement operation encoded by the message. The Shannon index assumes that the message is known with certainty once it has been received; if it is not, the Wiener index should be used.

---

[25]But anyway see the critiques of von Foerster (1960) and of Ashby (1962). We may, however, consider self-organization as programmable self-assembly.

[26]The creation of disorder could be avoided by doing things perfectly reversibly, but that implies doing them infinitely slowly and is, hence, scarcely of practical interest.

Information can be represented as a sign or as a succession of signs (symbols). The information conveyed by each symbol equals the freedom in choosing the symbol. If all choices are *a priori* equiprobable, the specification of a sequence removes uncertainty maximally. In practice, there may be strong syntactical constraints imposed on the successive choices, which limit the possible variety in a sequence of symbols.

In order to be considered valuable (or desired), the received information must be remembered (macroscopic information). Microinformation is not remembered. Thus, the information inherent in the positions and momenta of all the gas molecules in a room is forgotten picoseconds after its reception. It is of no value.

Information can be divided into three aspects: the signs themselves, their syntax (their relation with each other), and the accuracy with which they can be transmitted; their meaning, or semantic value (i.e., their relation to designata); and their effect (how effectively the received meaning affects the conduct of the recipient in the desired way), which may be called pragmatics, the study of signs in relation to their users, or significs, the study of significance.[27] In other words, content comprises the signs themselves and their syntax (i.e., the relation between them), their meaning (semantic value), and their effect on the conduct of the recipient (i.e., does it lead to action?). A further aspect is that of style, very difficult to quantify. It can be considered to be determined by word usage frequencies, from which the cybernetic temperature can be derived (cf. Eq. 3.7). An indication of style (cf. biomarkers giving an indication of disease) might be given by the occurrence of certain characteristic words, including the use of a certain synonym rather than another. If a symbolic sequence is modelled as a Markov chain, matters of style would be encapsulated in hidden Markov models (q.v.).

Meaning may be highly context-dependent; the stronger this dependence, the more implicit the meaning.

The effect of receipt of information on behaviour can be quantified in terms of changes to the logical pattern describing the awareness of the recipient to his environment. In simpler terms, this may be quantified as value in terms of a change in behaviour (assuming that enough data on replicate systems or past events are available to enable the course of action that would have taken place in the absence of the received information to be determined).

Information is inherently discrete (quantal) and thus based on combinatorics, which also happens to suit the spirit of the digital computer. In biology, if "genotype" constitutes the signs, then "phenotype" constitutes meaning. Action is self-explanatory and linked to adaptation (see Sect. 9.2). Biological function might be considered to be the potential for action.

---

[27] These three aspects, namely of syntactics, semantics, and pragmatics, are usually considered to constitute the theory of signs, or semiotics.

References

Ashby WR (1956) An introduction to cybernetics. Chapman & Hall, London

Ashby WR (1962) Principles of the self-organizing system. In: von Foerster H, Zopf GW (eds) Principles of self-organization. Pergamon Press, Oxford, pp 255–278

Bennett CH (1988) Logical depth and physical complexity. In: Herken R (ed) The Universal Turing Machine—A Half-Century Survey. Oxford University Press, Oxford, pp 227–257

Chernavsky DS (1990) Synergetics and information. Matematika Kibernetika 5:3–42 (in Russian)

Carnap R, Bar-Hillel Y (1952) An outline of a theory of semantic information. MIT Research Laboratory of Electronics Technical Report No 247

Dewey TG (1996) Algorithmic complexity of a protein. Phys Rev E 54:R39–R41

Dewey TG (1997) Algorithmic complexity and thermodynamics of sequence-structure relationships in proteins. Phys Rev E 56:4545–4552

Fisher RA (1951) The design of experiments, 6th edn. Oliver & Boyd, Edinburgh

von Foerster H (1960) On self-organizing systems and their environments. In: Yorvitz MC, Cameron S (eds) Self-organizing systems. Pergamon Press, Oxford

Good IJ (1969) Statistics of language. In: Meetham AR (ed) Encyclopaedia of linguistics, information and control. Pergamon Press, Oxford, pp 567–581

Karbowski J (2000) Fisher information and temporal correlations for spiking neurons with stochastic dynamics. Phys Rev E 61:4235–4252

Kullback S, Leibler RA (1951) On information and sufficiency. Ann Math Stat 22:79–86

Mackay DM (1950) Quantal aspects of scientific information. Phil Mag (ser 7) 41:289–311

Markov AA (1913) Statistical analysis of the text of "Eugene Onegin" illustrating the connexion with investigations into chains. Izv Imp Akad Nauk, Ser 6(3):153–162 (in Russian)

Shannon CE (1951) Prediction and entropy of printed English. Bell Syst Tech J 30:50–64

Thomas PJ (2010) An absolute scale for measuring the utility of money. J Phys: Conf Ser 238:012039

Tureck R (1995) Cells, functions, relationships in musical structure and performance. Proc R Inst 67:277–318

Welby V (1911) Significs. Encyclopaedia britannica, 11th edn. Cambridge University Press, Cambridge

Wiener N (1948) Cybernetics, or control and communication in the animal and the machine (Actualités Sci. Ind. no 1053). Hermann & Cie, Paris

Zurek WH (1989) Thermodynamic cost of computation, algorithmic complexity, and the information metric. Nature (Lond) 341:119–124

The Transmission of Information

3

In the previous chapter, although we spoke of the recipient of a message, implying also the existence of a dispatcher, the actual process of communicating between emitter and receiver remained rather shadowy. The purpose of this chapter is to explicitly consider transmission or communication channels.

Information theory grew up within the context of the transmission of messages and did not concern itself with appraisal of the meaning of a message. Later, Shannon (1951) (and others) went on to study the redundancy present in natural languages, since if the redundancy is taken into account in coding, the message can be compressed, and more information can be sent per unit time than would otherwise be possible (although, as we have noted in the previous chapter, much more compression may be achieved at the level of semantics or style).

Physically, channels can be extremely varied. The archetype used to be the copper telephone wire; nowadays, it would be an optical fibre. Consider the receipt of a weather forecast. A satellite orbiting the Earth emits an image of a mid-Atlantic cyclone or a remote weather station emits wind speed and temperature. Taking the first case, photons first had to fall on a detector array, initiating the flow of electrons along wires. These flows were converted into binary impulses (representing black or white; i.e., light or dark on the image) preceded by the binary address of each pixel. In turn, these electronic impulses were converted into electromagnetic radiation and beamed towards the Earth, where they were converted back into electrical pulses used to drive a printer, which produced an image of the cyclone on paper. This picture was viewed by the meteorologist, photons falling on his retina were converted into an internal representation of the cyclone in the meteorologist's brain, after some processing he composed a few sentences expounding the meaning of the information and its likely effect, these sentences were then spoken, involving the passage of neural impulses from brain to vocal chords, the sound emitted from his mouth travelled through the air, actuating resistance, hence electronic current fluctuations in a microphone, which travelled along a wire to be again converted into electromagnetic radiation, broadcast, and picked up by a wireless receiver, converted back into acoustic waves travelling through the air, picked up by the intricate mechanism of the ear, converted into nervous impulses, and processed by the brain of the listener. According to the

© Springer-Verlag London 2015
J. Ramsden, *Bioinformatics*, Computational Biology 21,
DOI 10.1007/978-1-4471-6702-0_3

nature of the message, muscles may then have been stimulated in order to make the listener run outside and secure objects from being blown away, or whatever. Perhaps, during the broadcast, some words may have been rendered unintelligible by unwanted interference (noise). It should also be mentioned that the whole process did not, of course, happen spontaneously, but the satellite and attendant infrastructure had previously been launched by the meteorologist with the specific purpose of providing images useful in weather forecasting. From this little anecdote we may gather that the transmission of information involves coding and decoding (i.e., transducing) messages, that transmission channels are highly varied physically, and that noise may degrade the information received.

Inside the living cell, it may be perceived that similar processes are operating. Sensors on the cell surface register a new carbon source, more abundant than the one on which the bacterium has been feeding, a conformational change in the sensor protein activates its enzymatic (phosphorylation) capability,[1] some proteins in its vicinity are phosphorylated, in consequence change conformation, and then bind to the promoter site for the gene of an enzyme able to metabolize the new food source. Messenger RNA is synthesized, templating the synthesis of the enzyme, which may be modified after translation. The protein folds to adopt a meaningful, enzymatically active structure and begins to metabolize the new food perfusing into the cell. Concomitant changes may result in the bacterium adopting a different shape—its phenotype changes.[2]

In very general terms, semiotics is the name given to the study of signals used for communication. In the previous chapter, the issues of the accuracy of signal transmission, the syntactical constraints reducing the variety of possible signals, the meaning of the signals (semantics), and their ultimate effect were broached. In this chapter we shall be mainly concerned with the technical question of transmission accuracy, although we shall see that syntactical constraints play an important rôle in considering channel capacity. We noted at the beginning of Chap. 2 that information theory has traditionally focused on the processes of transmission. In classical information theory, as exemplified by the work of Hartley (1928) and especially Shannon (1948), the main problem addressed is the capacity of a communication channel for error-free transmission. This problem was highly relevant to telegraph and telephone companies, but they were not in the least concerned with the nature of the messages being sent over their networks.

Some features involved in communication are shown in Fig. 3.1. There will always be a source (emitter), channel (transmission line), and sink (receiver), and encoding is necessary even in the simplest cases: For example, a man may say a sentence to a messenger who then runs off and repeats the message to the addressee, but of course to be able to do that he had to remember the sentence, which involved encoding the words as patterns of neural firing. Even if one simply speaks to an interlocutor and regards the mouth as the source, the mouth is not the receiver: The sounds are

---

[1]Section 14.7.
[2]These processes are considered in more detail in Chap. 10.

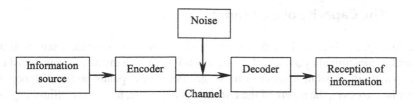

Fig. 3.1 Schematic diagram of subprocesses involved in transmitting a signal from a source to a receiver. Not all of the subprocesses shown are necessary, as discussed in the text. Noise may enter from the environment or may be intrinsic to the channel hardware

encoded as patterns of air waves and decoded via tiny mechanical movements in the ear.

What is the flow of information in the formal scheme of Fig. 3.1? In the previous chapter we essentially only considered one agent, who himself carried out an operation (such as measuring the length of a piece of wood), which reduced uncertainty and hence resulted in a gain of information according to Eq. (2.7) and further quantified by Eq. (2.5). We now consider that the information is encoded and transmitted (Fig. 3.2); indeed, it could be broadcast to an unlimited number of people. If they desired to know the length of that piece of wood and if the structure of their ignorance was the same as that of the measurer prior to the measurement (i.e., that the wood was less than a foot long, and they expected to receive the length in inches), then all those receiving that information would gain the same amount. The transmitted signals therefore have the potential for making a selection, by operating on the predefined set of alternatives, in exactly the same way as the actual act of measurement itself. The information content of signals is based on this potential for discrimination. Hartley (1928), in his pioneering paper, referred to the successive selection of signs from a given list. This is of course precisely what happens when sending a telegram.

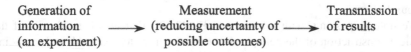

Fig. 3.2 Schematic diagram of the subprocesses involved in carrying out a physical experiment and transmitting the results

3.1   The Capacity of a Channel

Channel capacity is essentially dependent on the physical form of the channel. If the channel is constituted by a runner bearing a scroll on which the message is inscribed, the capacity, in terms of number of messages per day, depends on the distance the runner has to cover, the nature of the terrain, his physique, and so on.[3] Similarly, the capacity of a heliograph signalling system (in flashes per minute) depends on the dexterity of the operators working the mirrors and the availability of sunlight.

It is obviously convenient, when confronted with the practicalities of comparing the capacities of different channels (for example, a general in the field may have to decide whether to rely on runners or set up a heliograph) to have a common scale with which the capacities of different channels may be compared. A channel is essentially transmitting *variety*. A runner can clearly convey a great deal of variety, since he could bear a large number of different messages. If he can comfortably carry a sheet on which a thousand characters are written, and assuming that the characters are selected from the English alphabet plus space, then the variety of a single message is $1000 \log_2 27 = 4754$ bits to a first approximation. If the runner can convey three scrolls a day, the variety is then $3 \times 4754/(12 \times 3600) = 0.33$ bits per second, assuming 12 h of good daylight.

The heliograph operator, on the other hand, may be able to send one signal per second, with a linear variety of two (flash or no flash); that is, during the 12 h of good daylight, he can transmit with a rate of $\log_2 2 = 1$ bit/s.

It may be, of course, that the messages the general needs to send are highly stereotyped. Perhaps there are just 100 different messages that might need to be sent.[4] Hence, they could be listed and referred to by their number in the list. Since the number 100 (in base 10) can be encoded by $\log_2 100 = 6.64$ bits, any of the 100 messages could be sent within 7 s. Furthermore, if experience showed that only 10 of the messages were sent rather frequently (say with probability 0.05 each), and the remaining 90 with probability $\frac{0.5}{90}$, application of Eq. (2.5) shows that 5.92 bits would suffice, so that a more compact coding of the 100 messages could in principle be found.[5]

We note in passing, with reference to Eq. (2.13), that all of the details of the physical construction of the heliograph, or whatever system is used, and including the table of 100 messages assigning a number to each one, so only the number needs to be sent, are included in K. Should it be necessary to quantify K, it can be done via the algorithmic complexity (AIC; see Sect. 6.5), but as far as the transmission of messages is concerned, this is not necessary, since we are only concerned with the *gain* of information by the recipient (cf. Eq. 2.14).

---

[3]Note that here the information source is the brain of the originator of the message, and the encoder is the brain-hand-pen system that results in the message being written down on the scroll.

[4]Such stereotypy is extensively made use of in texting with a cell phone.

[5]Note that Shannon's theory does not give any clues as to how the most compact coding can be found.

The meaning of each message (i.e., an encoded number) sent under the second scheme could potentially be very great. It might refer to a book full of instructions. Here we shall not consider the effect of the message (cf. Sect. 2.3.3).

Another point to consider is possible interference with the message. The runner would be a target for the enemy; hence, it may be advisable to send, say, three runners in parallel with copies of the same message. It might also have been found that the distant heliograph operator had difficulty in receiving the flashes reliably from the sender, and it might therefore have been decided to repeat each flash three times and the recipient would use majority selection on each group of three to deduce the message. The capacity of the channel would thereby be lowered threefold.

In many practical cases, the physical medium for transmitting messages has to be shared by many different messages. It is a great advantage of optical communications that streams of photons of different wavelengths do not interfere with one another. Therefore, an optical fibre can carry many independent signals. Inside a cell, in which the cytoplasm is a shared medium, many different molecules are present and independence is determined by the differential chemical affinities between pairs of molecules.

3.2   Coding

Coding refers to the transduction of a message into another form. It is ubiquitous in our world. Ideas are encoded into words, music, pictures, one language may be encoded into another, and so on. We have already made extensive use of binary coding; the compact disk-based recording industry today uses binary coding almost exclusively for music, pictures and words. Evidently any number can be written in base 2; hence, a possible drill (algorithm) for binary coding consists of the following steps:

1. Assign a number to each state to be encoded;
2. Convert that number into base 2.

A DNA sequence can thereby be converted into binary form by making the assignments A $\to$ 1, C $\to$ 2, T $\to$ 3, and G $\to$ 4, which in base 2 are 1, 10, 11, and 100, respectively. The coded sequence would have to be written (001, 010, etc.) and read in groups of three digits, otherwise "AA" could be misinterpreted as "T" and so forth. Alternatively, separators can be introduced (see also the Huffman code described near the beginning of Sect. 3.4). The reading frame is thus defined as the series of groups of three beginning with the first. DNA is an example of a usually nonoverlapping code of contiguous triplets.

Codes may be written as transformations, e.g.,

$$\downarrow \begin{matrix} A & B & C & D & \cdots & Z \\ B & C & D & E & \cdots & A \end{matrix} \text{ ,}$$

which could also be written down compactly by the instruction "replace each letter by the next one to the right" (sfqmbdf fbdi mfuufs cz uif ofyu pof up uif sjhiu). A scheme for recoding DNA could be

$$\downarrow \begin{array}{cccc} A & C & T & G \\ 1 & 2 & 3 & 4 \end{array}$$

in any base above 4. As is well known, DNA is encoded by RNA using the transformation[6]

$$\downarrow \begin{array}{cccc} A & C & T & G \\ U & G & A & C \end{array}$$

by virtue of complementary base-pairing, and RNA triplets are, in turn, encoded by amino acids (Table 3.1).

Table 3.1 The genetic code

First (5′)	Second position				Third (3′)
	U	C	A	G	
U	phe	ser	tyr	cys	U
	phe	ser	tyr	cys	C
	leu	ser	stop	stop	A
	leu	ser	stop	trp	G
C	leu	pro	his	arg	U
	leu	pro	his	arg	C
	leu	pro	gln	arg	A
	leu	pro	gln	arg	G
A	ile	thr	asn	ser	U
	ile	thr	asn	ser	C
	ile	thr	lys	arg	A
	met	thr	lys	arg	G
G	val	ala	asp	gly	U
	val	ala	asp	gly	C
	val	ala	glu	gly	A
	val	ala	glu	gly	G

Note The table is given for RNA; for DNA, T must be used in place of U. See Table 11.6 for the key to the amino acid abbreviations. "stop" is an instruction to stop sequence translation. AUG encodes the corresponding instruction to "start" (in eukaryotes; sometimes other triplets are used in prokaryotes).

---

[6]Since DNA is composed of two complementary strands, one could equally well write the coding transformation as

$$\downarrow \begin{array}{cccc} A & C & T & G \\ A & C & U & G \end{array}.$$

Codes used in telecommunications are single-valued and one-to-one transformations (i.e., bijective functions), which allows unambiguous decoding. The type of coding found in biology is more akin to that described for the broadcast meteorological bulletin described at the beginning of this chapter, in which the physical carrier of the information changes and the bare technical content accrues meaning. In that example, supposing that the satellite was defined as the information source, the meteorologist could scarcely have made sense, in his head, of the stream of pixel densities, but as soon as they were interpreted by writing them down as black and white squares (which he could have done with pencil on paper had he been aware of the structure of the information, especially the order in which the pixels were to be arranged) it would have been apparent that they code for a picture; that is, there is a jump in meaning. So it is in biology—the amino acid sequence is structured in such a way that meaning is accrued, not only as a three-dimensional structure but as a functional enzyme or structural element, able to interact with other molecules.

Coding—signal transduction—is ubiquitous throughout the cell and between cells. Typically, a state of a cell is encoded as a particular concentration level of a small molecule (cf. Tomkins' 1975 "metabolic code"). For encoding this kind of information, a small number of small molecules, such as cyclic adenosine monophosphate (cAMP) and calcium ions (Ca^{2+}), is used. The chemical nature of these molecules is usually unrelated to the nature of the information they encode (see Chap. 16 for details).

3.3 Decoding

The main requirement for decoding in a transmission scheme is that the coding transformation is one-to-one and, hence, each encoded symbol has a unique inverse. In biological systems, decoding (in the sense of reconstituting the original message) may be relatively unimportant at the molecular level; the encoded message is typically used directly, without being decoded back into its original form as envisaged in Fig. 3.2.

The problem of decoding the simple transformations described in the previous section is straightforward. Consider now a scheme for encoding that uses a machine that can be in one of four states $\{A, B, C, D\}$ and that the transformation depends on an input parameter that can be one of $\{Q, R, S\}$. In tabular form,

$$
\begin{array}{c|cccc}
\downarrow & A & B & C & D \\
\hline
Q & D & A & B & C \\
R & C & D & A & B \\
S & B & C & D & A
\end{array}
\tag{3.1}
$$

Given an initial state, an input message in the form of a sequence of parameter values will result in a particular succession of states adopted by the machine; for example, if the machine (transducer) starts in state B, the parameter stream $QQSRQ$ will result in the subsequent output A, D, A, C, B. In tabular form,

$$\begin{array}{ll} \text{Input state:} & Q \quad Q \quad S \quad \dots \\ \text{Transducer state}: & B \quad A \quad D \quad A \end{array} \tag{3.2}$$

The problem faced by the decoder (inverter) is that although each transition gives unambiguous information about the parameter value under which it occurred, the two states involved did not exist at the same epoch; hence, one of the decoder's inputs must in effect behave *now* according to what the encoder's output *was*. This problem may be solved by introducing a delayer, represented by the transformation

$$\begin{array}{c|ccc} \downarrow & q & r & s \\ \hline Q & q & q & q \\ R & r & r & r \\ S & s & s & s \end{array} \tag{3.3}$$

The encoder provides input (is joined) to the delayer and the decoder, and the delayer provides an additional input (is joined) to the decoder (see the following example).

Example. Consider a transducer (encoder) with the transformation $n' = n + a$, where a is the input parameter and n is the variable.[7] The inverting solution of the transducer's equation is evidently $a = n' - n$, but since n' and n are not available simultaneously, a delayer is required. The delayer should have the transformation $n' = p$, with n as the parameter and p as the variable. The inverter (decoder) has variable m and inputs n and p, and its transformation is $m' = n - p$. The encoder's input to the delayer and the decoder is n, and the delayer's to the decoder is its state p.

Problem. Start the transducer in the above example with $n = 3$ and verify the coding–decoding operation.

Problem. Attempt to find examples of decoders in living organisms.

3.4   Compression

Shannon's fundamental theorem for a noiseless channel proves that it is possible to encode the output of an information source in such a way as to transmit at an average rate equal to the channel capacity.

This is of considerable importance in telephony, which mostly deals with the transmission of natural language. Shannon found by an empirical method that the redundancy of the English language (due to syntactical constraint) is about 0.5. Hence, by suitably encoding the output of an English-speaking source, the capacity of a channel may be effectively doubled.

This compression process is well illustrated by an example due to Shannon. Consider a source producing a sequence of letters chosen from among A, B, C, and D. Our

---

[7]Due to Ashby (1956).

first guess would be that the four symbols were being chosen with equal probabilities of $\frac{1}{4}$, and hence the average information rate per symbol would be $\log_2 4 = 2$ bits per symbol. However, suppose that after a long delay we ascertain from the frequencies that the probabilities are respectively $\frac{1}{2}$, $\frac{1}{4}$, $\frac{1}{8}$, and $\frac{1}{8}$. Then, from Eq. (2.5) we determine $I = 1.75$ bits per symbol, so we should be able to encode the message (whose relative entropy is $\frac{7}{8}$ and hence redundancy R is $\frac{1}{8}$) such that a smaller channel will suffice to send it. The following code may be used:[8]

$$
\begin{array}{cccc}
& A & B & C & D \\
\downarrow & 0 & 10 & 110 & 111
\end{array} \, .
$$

The average number of binary digits used in encoding a sequence of N symbols will be $N(\frac{1}{2} \times 1 + \frac{1}{4} \times 2 + \frac{2}{8} \times 3) = \frac{7}{4}N$. 0 and 1 can be seen to have equal probabilities; hence, I for the coded sequence is 1 bit/symbol, equivalent to 1.75 binary symbols per original letter. The binary sequence can be decoded by the transformation

$$
\begin{array}{cccc}
& 00 & 01 & 10 & 11 \\
\downarrow & A' & B' & C' & D'
\end{array}
$$

The compression ratio of this process is $\frac{7}{8}$. Note, however, that there is no general method for finding the optimal coding.

Problem. Using the above coding, show that the 16-letter message "ABBAAAD-ABACCDAAB" can be sent using only 14 letters.

The Shannon technique requires a long delay between receiving symbols for encoding and the actual encoding, in order to accumulate sufficiently accurate individual symbol transmission probabilities. The entire message is then encoded. This is, of course, a highly impractical procedure. Mandelbrot (1952) has devised a procedure whereby messages are encoded word by word. In this case the word delimiters (e.g., spaces in English text) play a crucial rôle. From Shannon's viewpoint, such a code is necessarily redundant, but on the other hand, an error in a single word renders only that word unintelligible, not the whole message. It also avoids the necessity for a long delay before coding can begin.

The Mandelbrot coding scheme has interesting statistical properties. One may presume that the encoder seeks to minimize the cost of conveying a certain amount of information using the collection of words that are at his disposal. If p_i is the probability of selecting and transmitting the ith word, then the mean information per symbol contained in the message is, as before, $-\sum p_i \log p_i$. We may suppose that the cost of transmitting a selected word is proportional to its length. If c_i is the cost of transmitting the ith word, then the average cost per word is $\sum p_i c_i$. Minimizing the distribution of the probabilities while keeping the total information constant (using Lagrange's method of undetermined multipliers) yields

$$
p_i = C e^{-D c_i} , \tag{3.4}
$$

---

[8]Elaborated by D.A. Huffman.

a sort of Boltzmann distribution. C is a constant fixed by the condition that $\sum p_i = 1$, and D is an as yet undetermined constant.

Suppose that the words are made up of individual letters (symbols) and demarcated by a special word demarcation symbol (the space in many languages). Cost, length, and number of letters are all proportional to each other. If the letters can be chosen in any way from an alphabet of A different ones, by the multiplication rule (Sect. 4.2.1) there are A^n different n-letter words. Let these words now be ranked in order of increasing cost and call this rank r. Since the cost increases linearly with n, it only increases logarithmically with rank,[9] that is,

$$c_r = \log_A r . \tag{3.5}$$

Substituting Eq. (3.5) into (3.4), one obtains a power law relation

$$p_r = Cr^{-B} , \tag{3.6}$$

known as Zipf's law when $B = 1$. Mandelbrot has shown that, more precisely, Eq. (3.6) is

$$p_r = C(r + \rho)^{-B} \tag{3.7}$$

and that the constant B (subsuming D in Eq. 3.4), the reciprocal of the informational temperature θ of the distribution (by analogy with the thermodynamic case), can take values other than 1. For $B > 1$ (i.e., $\theta < 1$) the language is called open (because the value of C does not greatly depend on the total number of words), whereas for $B < 1$ it does, and the corresponding language is called closed. The constant ρ is connected with the freedom of choosing words (cf. Sect. 4.2.3), but a deep interpretation of its significance in messages has not yet been given. Equation (3.7) fits the distribution of written texts remarkably well, and most languages such as English, German, and so forth are open, whereas highly stylized languages (e.g., modern Hebrew and the English of the Pennsylvania Dutch) are closed. θ is a measure of the agility of exploiting vocabulary; low values are characteristic of children learning a language or schizophrenic adults; the richest and most imaginative use of vocabulary corresponds to $\theta = 1$.

There are many heuristic methods for compression. Dictionaries (i.e., lists of frequent words) are often used for word texts. In rastered images, successive lines typically show small changes; large blocks are uniformly black, grey or white, and so on. A useful way of compressing long sequences of symbols is to search for segments that are duplicated. The duplicates can then be encoded by the distance of the match from the original sequence and the length of the matching sequence (number of symbols). Zipping software typically works on this principle;[10] the compression is greatest for files with a lot of repetitive material, but according to van der Waerden's (1927) extension of Baudet's conjecture, any string of two kinds of symbols has repetitive sequences of at least one of the symbols.

---

[9]The words are listed in order of increasing cost; rank 1 has the lowest cost and so on.
[10]e.g., Ziv and Lempel (1977).

3.4.1 Use of Compression to Measure Distance

Suppose two ergodic binary sources P and Q emit 1 s with probabilities p and q, respectively. The Kullback and Leibler (1951) relative entropy between the two strings is

$$S_{PQ} = -q \log_2 \frac{p}{q} - (1-q) \log_2 \frac{1-p}{1-q} \tag{3.8}$$

and may be used as the basis of a measure of distance between the two strings. Benedetto et al. (2002) have devised an ingenious method for estimating S_{PQ} from two sources by zipping a long string from each source (P and Q) and the same long strings to each of which is appended a sufficiently short string fragment (say P') from one of the sources. S_{PQ} is then the difference in coding efficiency between P' coded optimally because it follows P (the source is ergodic) and P' coded suboptimally because it follows Q. Using L to denote the length of a zipped file,

$$S_{PQ} = [(L_{Q+P'} - L_Q) - (L_{P+P'} - L_P)]/L'_{P'} \tag{3.9}$$

(in bits per character), where $L'_{P'}$ is the unzipped length of the short string fragment P'. In order to eliminate dependency on the particular coding, a different normalization may be used:

$$S_{PQ} = \frac{(L_{Q+P'} - L_Q) - (L_{P+P'} - L_P)}{L_{P+P'}} + \frac{(L_{P+Q'} - L_P) - (L_{Q+Q'} - L_Q)}{L_{Q+Q'}}. \tag{3.10}$$

3.4.2 Ergodicity

Ergodicity means that every allowable point in phase space is visited infinitely often in infinite time or, in practice, every allowable point in phase space is approached arbitrarily closely after a long time. Ergodicity is a pillar of Boltzmann's assumption that the microstates of an ensemble have equal *a priori* probabilities, and indeed of the rest of statistical mechanics. Nevertheless, as our knowledge of the world has increased, it has become apparent that ergodicity actually applies only to a small minority of natural systems. Although some systems may not even be ergodic in the infinite time limit, most observed departures from ergodicity occur because of the inordinately long times that would be required to fulfil it. The departures are particularly common in condensed matter: any glass, for example, breaks ergodicity. In nonergodic systems, the phase space or ensemble average does not equal the time average.

A homely illustration of some of the issues to be considered, in particular that breaking ergodicity depends on the timescale of the observer, is provided by a cup of hot coffee to which cream is added and stirred. The coffee and cream become homogeneously mixed within seconds, the cup and contents reach the temperature of the surroundings after tens of minutes, and the water evaporates and is in equilibrium with the atmosphere in the room after several hours. Whether the observed behaviour is representative of the allowed phase space depends on the observational timescale τ_0. In general, broken ergodicity can be expected if there are significant dynamical timescales longer than τ_0.

In a more general sense, applicable also to symbolic strings, ergodic means that any one exemplar (substring) is typical of the ensemble; hence, if the string is ergodic, it is to be expected that every permissible sequence will be encountered. Clearly, therefore, the DNA of living organisms is not ergodic (although it might be argued that hitherto we have taken a too liberal view of what is "permissible").

3.5   Noise

So far we have supposed that the messages received over the communication channel are precisely those transmitted. This is a rather idealized situation. We have doubtlessly had the experience of speaking on a very noisy telephone line, or listening to a radio with very poor reception, and only been able to make out one word in two perhaps, and yet could still understand what was being said. The syntactical redundancy of English is about 0.5; hence, it is not surprising that about half the words or symbols may be removed (at random) without overly impairing our ability to receive the original message.

According to our previous discussion of the Shannon index, I is additive for independent sources of uncertainty. Noise is an independent source of uncertainty and can be treated within the theoretical framework we have discussed.

Suppose that signal x was sent and y was received, the difference between the two being due to noise. The amount of information lost in transmission is called the equivocation, E.

Definition. The equivocation is

$$E = I(x) - I(y) + I_x(y) , \tag{3.11}$$

where $I(x)$ is the information sent, $I(y)$ is the information received, and $I_x(y)$ is the uncertainty in what was received if the signal sent be known.[11]

The concept of equivocation enables one to write the actual rate of information transmission $\mathcal{R}$ over a noisy channel in a rather transparent way:

$$\mathcal{R} = I(x) - E ; \tag{3.12}$$

that is, the rate equals the rate of transmission of the original signal minus the uncertainty in what was sent when the message received is known. From our definition (3.11),

$$\mathcal{R} = I(y) - I_x(y) , \tag{3.13}$$

---

[11]It should be clear that the information sent is already the result of some measurement operation or whatever, in the sense of our previous discussion.

where $I_x(y)$ is the spurious part of the information received (i.e., the part due to noise) or, equivalently, the average uncertainty in a message received when the signal sent is known. It follows (cf. Sect. 4.1) that

$$\mathcal{R} = I(x) + I(y) - I(x, y) , \tag{3.14}$$

where $I(x, y)$ is the joint entropy of input (information transmitted) and output (information received). By symmetry, the joint entropy equals

$$I(x, y) = I(x) - I_x(y) = I(y) - I_y(x) . \tag{3.15}$$

We could just as well write E as $I_y(x)$: it is the uncertainty in what was sent when it is known what was received. If there is no noise, $I(y) = I(x)$ and $E = 0$.

Let the error rate be η per symbol. Then

$$E = I_y(x) = \eta \log \eta + (1 - \eta) \log(1 - \eta) . \tag{3.16}$$

The maximum error rate is 0.5 for a binary transmission; the equivocation is then 1 bit/symbol and the rate of information transmission is zero.

The equivocation is just the conditional or relative entropy and can also be derived using conditional probabilities. Let $p(i)$ be the probability of the ith symbol being transmitted and let $p(j)$ be the probability of the jth symbol being received. $p(j|i)$ is the conditional probability of the jth signal being received when the ith was transmitted, $p(i|j)$ is the conditional probability of the ith signal being transmitted when the jth was received (posterior probability), and $p(i, j)$ is the joint probability of the ith signal being transmitted and the jth received.

The ignorance removed by the arrival of one symbol is (cf. Eq. 2.7)

$$I = \text{initial uncertainty} - \text{final uncertainty}$$

$$= \log p(i) - (-\log p(j))$$

$$= \log \frac{p(i|j)}{p(i)} . \tag{3.17}$$

Averaging over all i and j,

$$\bar{I} = \sum_i \sum_j p(i, j) \log \frac{p(i|j)}{p(i)} , \tag{3.18}$$

but since $p(i, j) = p(i)p(j|i) = p(j)p(i|j)$ (cf. Sect. 5.2.2),

$$\bar{I} = \sum_i \sum_j p(i, j) \log \frac{p(i, j)}{p(i)p(j)} . \tag{3.19}$$

If $i = j$ always, then we recover the Shannon index (Eq. 2.5). If the two are statistically independent, $\bar{I} = 0$.

From our definition of $p(i, j)$ we can write the posterior probability as

$$p(i, j) = \frac{p(i)}{p(j)} p(j, i) . \tag{3.20}$$

Shannon's fundamental theorem for a discrete channel with noise proves that if the channel capacity is C and the source transmission rate is $\mathcal{R}$, then if $\mathcal{R} \leq C$, there exists a coding system such that the source output can be transmitted through the channel with an arbitrarily small frequency of errors. The capacity of a noisy channel is defined as

$$C_{\text{noisy}} = \max(I(x) - E) ,\tag{3.21}$$

the maximization being over all sources that might be used as input to the channel.

3.6   Error Correction

Suppose a binary transmission channel had a 20 % chance of transmitting an incorrect signal; hence, a message sent as "0110101110" might appear as "1100101110". An easy way to render the system immune from such noise would be to repeat each signal threefold and incorporate a majority detector in the receiver. Hence, the signal would be sent as "000111111000111000111111111000" and received as "001011011000110000101111111100" (say), but majority detection would still enable the signal to be correctly restored. The penalty, of course, is that the channel capacity is reduced to a third of its previous value.

Many physical devices are so designed to be immune, to a certain degree, to random fluctuations in the physical quantities encoding information. In a digital device, zero voltage applied to a terminal represents the digit "0", and 1 V (say) represents the digit "1". In practice, any voltage up to about 0.5 will be interpreted as zero, and all voltages above 0.5 will be interpreted as 1.0 (see Fig. 3.3).

It is perfectly possible to devise codes that can detect and correct errors. Hamming defines systematic codes as those in which each code symbol has exactly n binary digits, m being associated with the information being conveyed and $k = n - m$ being used for error detection and correction. The redundancy (cf. Eq. 2.18) of a systematic code (s.c.) is defined as

$$R_{\text{s.c.}} = n/m .\tag{3.22}$$

Hamming (1950) constructed a single error-detecting code as follows: Information is placed in the first $n - 1$ positions of n binary digits. Either a 0 or a 1 is placed in the nth position, the choice being made to ensure an even number of 1 s in the n

Fig. 3.3 Input–output relationships for a device such as an electromechanical relay (*solid line*) and a field-effect transistor (*dashed line*)

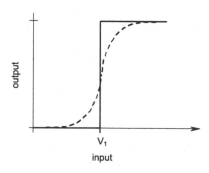

output

V_1

input

digit word. A single (or odd number of) error would leave an odd number of 1s in the word. Clearly, the redundancy is $n/(n-1)$. This type of error-detecting code is called a parity check; this particular one is an even parity check. n should be small enough such that the probability of more than one error is negligible.

To make an error-correcting code, a larger number ($k > 1$) of positions are given to parity checking and filled with values appropriate to selected information positions. When the message is received, k checks are applied in order, and if the observed value agrees with the previously calculated value, one writes a 0, but a 1 if it disagrees, in a new number called the checking number, which must give the position of any single error—i.e., it must describe $m + k + 1$ different things—hence, k must satisfy

$$m + k + 1 \leq 2^k \leq 2^n/(n+1) .\qquad(3.23)$$

The principle can obviously be extended to double error-correcting codes, which, of course, further increase the redundancy.[12]

3.7 Summary

Messages may be encoded in order to send them along a communication channel. Shannon's fundamental theorem proves that a message with redundancy can always be encoded to take advantage of it, enabling a channel to transmit information up to its maximum capacity.

The capacity of a channel is the number of symbols m that can be transmitted in unit time multiplied by the average information per symbol:

$$C = m\bar{I} .\qquad(3.24)$$

Any strategy for compressing a message is actually a search for regularities in the message, and thus compression of transmitted information actually lies at the heart of general scientific endeavour.

Noise added to a transmission introduces equivocation, but it is possible to transmit information through a noisy channel with an arbitrarily small probability of error, at the cost of lowering the channel capacity. This introduces redundancy, defined as the quotient of the actual number of bits to the minimum number of bits necessary to convey the information. Redundancy therefore opposes equivocation; that is, it enables noise to be overcome. Many natural languages have considerable redundancy. Technical redundancy arises through syntactical constraints. The degree of semantic redundancy of English, or indeed of any other language, is currently unknown.

Problem. Attempt to define, operationally or otherwise, the terms "message", "message content", and "message structure".

---

[12]See also Levenshtein (2001) on the problem of efficient reconstruction of an unknown sequence from versions distorted by noise.

Problem. Calculate the amount of information in a string of DNA coding for a protein. Repeat for the corresponding messenger RNA and amino acid sequences. Is the latter the same as the information contained in the final folded protein molecule?

Problem. Discuss approaches to the problem of determining the minimum quantity of information necessary to encode the specification of an organ.

Problem. Is it useful to have a special term "bioinformation"? What would its attributes be?

References

Ashby WR (1956) An introduction to cybernetics. Chapman & Hall, London
Benedetto D, Caglioti E, Loreto V (2002) Language trees and zipping. Phys Rev Lett 88:048702
Hamming RW (1950) Error detecting and error correcting codes. Bell Syst Tech J 26:147–160
Hartley RVL (1928) Transmission of information. Bell Syst Tech J 7:535–563
Kullback S, Leibler RA (1951) On information and sufficiency. Ann Math Stat 22:79–86
Levenshtein VI (2001) Efficient reconstruction of sequences. IEEE Trans Inf Theory 47:2–22
Mandelbrot B (1952) Contribution à la théorie mathématique des jeux de communication. Publ Inst Statist Univ Paris 2:1–124
Shannon CE (1948) A mathematical theory of communication. Bell Syst Tech J 27:379–423
Shannon CE (1951) Prediction and entropy of printed English. Bell Syst Tech J 30:50–64
Tomkins GM (1975) The metabolic code. Science 189:760–763
van der Waerden BL (1927) Beweis einer Baudet'schen Vermutung. Nieuw Arch. Wiskunde 15:212–216
Ziv J, Lempel A (1977) A universal algorithm for sequential data compression. IEEE Trans Inf Theory IT–23:337–343

Sets and Combinatorics

4

4.1 The Notion of Set

Set is a fundamental, abstract notion. A set is defined as a collection of objects, which are called the *elements* or *points* of the set. The notions of union ($A \cup B$, where A and B are each sets), intersection ($A \cap B$), and complement (A^c) correspond to everyday usage. Thus, if $A = \{a, b\}$ and $B = \{b, c\}$, $A \cup B = \{a, b, c\}$, $A \cap B = \{b\}$, and $A^c = \{c, d, \ldots, z\}$ if our world is the English alphabet. *Functions* can be thought of as operations that map one set onto another.

Typically, all the elements of a set are of the same type; for example, a set called "apples" may contain apples of many different varieties, differing in their colours and sizes, but no oranges or mangoes; a set called "fruit" could, however, contain all of these, but no meat or cheese.

One is often presented with the problem of finding or estimating the size of a set. Size is the most basic attribute, even more basic than the types of elements. If the set is small, the elements can be counted directly, but this quickly becomes tedious and, as the set becomes large, it may be unnecessary to know the exact size. Hence, computational short cuts have been developed, which are usually labelled combinatorics. Combinatorial problems are often solved by looking at them in just the right way and, at an advanced level, problems tend to be solved by clever tricks rather than the application of general principles.

Problem. Draw Venn diagrams corresponding to $\cap$, $\cup$ and complement.

4.2 Combinatorics

Most counting problems can be cast in the form of making selections, of which there are four basic types, corresponding to with or without replacement, each with or without ordering. This is equivalent to assembling a collection of balls by taking them from boxes containing different kinds of balls.

© Springer-Verlag London 2015

J. Ramsden, *Bioinformatics*, Computational Biology 21,

DOI 10.1007/978-1-4471-6702-0_4

The Basic Rule of Multiplication

Consider an ordered r-tuple $(a_1, \ldots, a_r)$, in which each member a_i belongs to a set with n_i elements. The total number of possible selections equals $n_1 n_2 \cdots n_r$; for example, we select r balls, one from each of r boxes, where the ith box contains n_i different balls.

4.2.1 Ordered Sampling with Replacement

If all the sets from which successive selections are taken are the same size n, the total number of ordered (distinguishable) selections of r objects from n with repetition (replacement) allowed follows from the multiplication rule

$$\prod_i^r n_i = n^r \ . \tag{4.1}$$

In terms of putting balls in a row of cells, this is equivalent to filling r consecutive cells with n possible choices of balls for each one; after taking a ball from a central reservoir, it is replenished with an identical ball.

4.2.2 Ordered Sampling Without Replacement

If the balls are not replenished after removal, there are only $(n-1)$ choices of ball for filling the second cell, $(n-2)$ for the third, and so on. If the number of cells equals the number of balls (i.e., $r = n$), then there are $n!$ different arrangements—this is called a permutation (and can be thought of as a bijective mapping of a set onto itself); more generally, if $r \leq n$, the number of arrangements is

$$^nP_r = n(n-1)\cdots(n-r+1) = \frac{n!}{(n-r)!} \ , \tag{4.2}$$

remembering that 0! is defined as being equal to 1.

Random Choice

This means that all choices are equally probable. For random samples of fixed size, all possible samples have the same probability n^{-r} with replacement and $1/^nP_r$ without replacement. The probability of no repetition in a sample is therefore given by the ratio of these probabilities: $^nP_r/n^r$. Criteria for randomness are dealt with in detail in Chap. 6.

Stirling's Formula

This is useful for (remarkably accurate) approximations to $n!$:

$$n! \sim (2\pi)^{\frac{1}{2}} n^{(n+\frac{1}{2})} e^{-n} \ . \tag{4.3}$$

A simpler, less accurate, but easier to remember formula is

$$\log n! \sim n \log n - n \ . \tag{4.4}$$

4.2.3 Unordered Sampling Without Replacement

Suppose now that we repeat the operation carried out in the previous subsection, but without regard to the order; that is, we simply select r elements from a total of n. Let W be the number of ways in which it can be done. After having made the selection, we then order the elements, to arrive at the result of the previous subsection; that is, each selection can be permuted in $r!$ different ways. These two operations give us the following equation:

$$\frac{n!}{(n-r)!} = Wr! \tag{4.5}$$

The expression for W, the number of combinations of r objects out of n, which we shall now write as $^{n}C_{r}$ or $\binom{n}{r}$, follows immediately:

$$^{n}C_{r} = \binom{n}{r} = \frac{n!}{r!(n-r)!}, \tag{4.6}$$

with $\binom{n}{0} = 1$ from the definition of $0! = 1$. This is equivalent to stating that a population of n elements has $\binom{n}{r}$ different subpopulations of size $r \leq n$. Note that

$$\binom{n}{r} = \binom{n}{n-r} \quad \text{for } r = 0, 1, ..., n ; \tag{4.7}$$

in words, for example, selecting five objects out of nine is the same as selecting four to be omitted.

It is implied that the selections are independent. In practical problems, this may be far from reality. For example, a manufacturer assembling engines from 500 parts may have to choose from a total of 9000. The number of combinations is at first sight a huge number, $9000!/(500!\,8500!) \sim 10^{840}$ by Stirling's approximation, posing a horrendous logistics problem. Yet many of the choices will fix others; strong constraints drastically reduce the freedom of choice of the components.

Partitioning

The number of ways in which n elements can be partitioned into k subpopulations, the first containing r_1 elements, the second r_2, and so on, where $r_1 + r_2 + \cdots + r_k = n$, is given by multinomial coefficients $n!/(r_1!r_2!\cdots r_k!)$, obtained by repeated application of Eq. (4.6). If r balls are placed in n cells with occupancy numbers $r_1, r_2, \ldots, r_n$, with all n^r possible placements equally possible, then the probability to obtain a set of given occupancy numbers equals $n^{-r}n!/(r_1!r_2!\ldots r_k!)$ (the Maxwell-Boltzmann distribution). This multinomial coefficient will be denoted using square brackets:

$$\begin{bmatrix} r \\ r_i \end{bmatrix} = \frac{n!}{r_1!r_2!\cdots r_k!}, \quad \text{with } \sum_{i}^{i=k} r_i = n . \tag{4.8}$$

Fermi-Dirac Statistics

Fermi-Dirac statistics are based on the following hypotheses: (i) No more than one element can be in any given cell (hence $r \leq n$) and (ii) all distinguishable arrangements satisfying (i) have equal probabilities.

By virtue of (i), an arrangement is completely specified by stating which of the n cells contain an element; since there are r elements, the filled cells can be chosen in $\binom{n}{r}$ ways, each with probability $\binom{n}{r}^{-1}$.

Bose-Einstein Statistics

Let the occupancy numbers of the cells be given by

$$r_1 + r_2 + \cdots + r_n = r . \tag{4.9}$$

The number of distinguishable distributions (if the elements are indistinguishable, distributions are distinguishable only if the corresponding n-tuples $(r_1, \ldots, r_n)$ are not identical) is the number of different solutions of Eq. (4.9). We call this $A_{r,n}$ (given by Eq. 4.11) and each solution has the probability $A_{r,n}^{-1}$ of occurring.

Problem. Consider a sequence of two kinds of elements: a alphas, numbered 1 to a, and b betas numbered $a + 1$ to $a + b$. Show that the alphas and betas can be arranged in exactly

$$\frac{(a+b)!}{a!b!} = \binom{a+b}{a} = \binom{a+b}{b}$$

distinguishable ways.

4.2.4 Unordered Sampling with Replacement

This last of the four basic selection possibilities is exemplified by throwing r dice (i.e., placing r balls into $n = 6$ cells). The event is completely described by the occupancy numbers of the cells; for example, 3, 1, 0, 0, 0, 4 represents three 1s, one 2, and four 6s.

Generalizing, every n-tuple of integers satisfying

$$r_1 + r_2 + \cdots + r_n = r \tag{4.10}$$

describes a possible configuration of occupancy numbers. Let the n cells be represented by the n spaces between $n+1$ bars. Let each object in a cell be represented by a star (for the example given above, the representation would be $|\; {*}{*}{*} \;|\; {*} \;|||| \; {*}{*}{*}{*} \;|$). The sequence of stars and bars starts and ends with a bar, but the remaining $n - 1$ bars and the r elements placed in the cells can appear in any order. Hence, the number of distinguishable distributions $A_{r,n}$ equals the number of ways of selecting r places out of $n - 1 + r$ symbols. From Eq. (4.6) this is

$$A_{r,n} = \binom{n-1+r}{r} = \binom{n-1+r}{n-1} . \tag{4.11}$$

If we impose a condition that no cell be empty, the r stars leave $r - 1$ spaces, of which $n - 1$ are to be occupied by bars; hence, there are $\binom{r-1}{n-1}$ choices.

Problem. How many different DNA hexamers are there? How many different hexapeptides are there?

Problem. Estimate the fraction of actual DNA sequences (i.e., the genomes of known species) compared with all possible DNA sequences. Clearly state all assumptions.

4.3 The Binomial Theorem

Newton's binomial formula,

$$(a+b)^n = \sum_{k=0}^{n} \binom{n}{k} a^k b^{n-k} , \tag{4.12}$$

where a and b can also be compound expressions, can be derived by combinatorial reasoning; for example, $(a+b)^5 = (a+b)(a+b)(a+b)(a+b)(a+b)$, and to generate the terms, an a or b is chosen from each of the five factors.

Problem. Generalize the binomial theorem by replacing the binomial $a+b$ by a multinomial $a_1 + a_2 + \cdots + a_r$.

Probability and Likelihood

5

5.1 The Notion of Probability

In everyday speech, statements such as "probably the train will be late" or "probably it will be foggy tomorrow" have the character of judgements. Formally, however (i.e., in the sense used throughout this book), probabilities do not refer to judgments, but to *possible results (outcomes) of an experiment*. These outcomes constitute the "sample space".[1] For example, attributing a probability of 0.6 to an event means that the event is expected to occur 60 times out of 100. This is the "frequentist" concept of probability, based on *random choices from a defined population*.

The frequentist concept is sometimes called the "objective" school of thought: The probability of an event is regarded as an objective property of the event (which has occurred), measurable via the frequency ratios in an actual experiment. Historically, it has been opposed by the "subjective" school,[2] which regards probabilities as expressions of human ignorance; the probability of an event merely formalizes the feeling that an event will occur, based on whatever information is available.[3] The purpose of theory is then merely to help in reaching a plausible conclusion when there is not enough information to enable a certain conclusion to be reached. A pillar of this school is Laplace's *Principle of Insufficient Reason*: Two events are to be assigned equal probabilities if there is no reason to think otherwise. Under such circumstances, if information were really lacking, the objectivist would refrain from attempting to assign a probability.

These differing schools have a bearing on the whole concept of causality, and it may be useful to recall here some remarks of Max Planck (1932).[4] One starts with the

---

[1] Called *Merkmalraum* ("label space") in R. von Mises' (1931) treatise *Wahrscheinlichkeitsrechnung*.
[2] Its protagonists include Laplace, Keynes, and Jeffreys.
[3] According to J.M. Keynes, probability is to be regarded as "the degree of our rational belief in a proposition".
[4] Made during the 17th Guthrie Lecture to the Physical Society in London.

© Springer-Verlag London 2015

J. Ramsden, *Bioinformatics*, Computational Biology 21,

DOI 10.1007/978-1-4471-6702-0_5

proposition that a necessary condition for an event to be causally conditioned is that it can be predicted with certainty. If, however, we compare a prediction of a physical phenomenon with more and more accurate measurements of that phenomenon, one is forced to reach a remarkable conclusion—that in not a single instance is it possible to predict a physical event exactly, unlike a purely mathematical calculation. The "indeterminists" interpret this state of affairs by abandoning strict causality and asserting that every physical law is of a statistical nature. The opposing school of "determinists" asserts that the laws of nature apply to an idealized world-picture, in which phenomena are represented by precise mathematical symbols, which can be operated on according to strict and generally agreed rules and to which precise numbers can be assigned (to which an actual measurement can only approximate); in the corresponding mentally constructed world-picture, all events follow definable laws and are strictly determined causally; the uncertainty in the prediction of an event in the world of sense is due to the uncertainty in the translation of the event from the world of sense to the world-picture and vice versa. It is left to the interested reader to pursue the implications with respect to quantum mechanics (with which we will not be explicitly concerned in this book).

Sommerhoff (1950) formulated probability in the following terms: Given a system whose initial state can be one of a set Q of n alternatives $Q_1, Q_2, \ldots, Q_n$, of which a certain fraction m/n will lead to the subsequent occurrence of an event E that is to be expected in the normal development of the system, then the probability that any particular member of Q leads to E is given by the fraction m/n. Note that this formulation only applies to the effects of the initial states, not to the states themselves. It has the advantage of avoiding any assumption of equally probable, or equally uncertain, events.

Before any further discussion about probability can take place, it is essential to agree on what is meant by the *possible results from an experiment (or observation)*. These results are called "events". Very often abstract models, corresponding to idealized events, are constructed to assist in the analysis of a phenomenon.

5.2   Fundamentals

The elementary unit in probability theory is the *event*. One has a fair freedom to define the event; simple events are irreducible and compound events are combinations of simple events. For example, the throw of a die to produce a 5 (with probability 1/6) is a simple event, and combinations of events to yield the same final result, such as three 2s, or a 5 and a 1, are compound events. Implicitly, the level of description is fixed when speaking of events in this way; clearly, the "event" of throwing a 6 requires many "sub-events" (which are events in their own right) involving muscular movements and nervous impulses, but these take place on a different level.

The general approach to solving a problem requiring probability is as follows:

1. Choose a set to represent the possible outcomes;
2. Allocate probabilities to these possible outcomes.

The results of probability theory can be derived from three basic axioms, referring to events and their totality in a manner that we must take to be carefully circumscribed:[5]

$$P\{E\} \geq 0 \text{ for every event } E \,, \tag{5.1}$$

$$P\{S\} = 1 \text{ for the certain event } S \,, \tag{5.2}$$

$$P\{A\} = \sum_i P\{a_i\} \,. \tag{5.3}$$

S includes all possible outcomes. Hence, if E and F are mutually exclusive events, the probability of their joint occurrence (corresponding to the AND relation in logic; i.e., "E and F") is simply the sum of their probabilities:

$$P\{E \cup F\} = P\{E\} + P\{F\} \,. \tag{5.4}$$

Simple events are by definition mutually exclusive ($P\{E\} \cap P\{F\} = 0$), but compound events may include some simple events that belong to other compound events and, more generally (inclusive OR; i.e., "E or F or both"),

$$P\{E \cup F\} = P\{E\} + P\{F\} - P\{EF\} \,. \tag{5.5}$$

If events are independent, then the probability of occurrence of those portions shared by both is

$$P\{E \cap F\} = P\{EF\} = P\{E\}P\{F\} \,. \tag{5.6}$$

It follows that for equally likely outcomes (such as the possible results from throwing a die or selecting from a pack of cards), the probabilities of compound events are proportional to the numbers of equally probable simple events that they contain:

$$P\{A\} = \frac{N\{A\}}{N\{S\}} \,. \tag{5.7}$$

We used this result at the beginning of this section to deduce that the probability of obtaining a 5 from the throw of a die is 1/6.

Problem. Prove Eqs. (5.4) and (5.5) with the help of Venn diagrams.

---

[5]Notation: in this chapter, $P\{X\}$ denotes the probability of event X; $N\{X\}$ is the number of simple events in (compound) event X. S denotes the certain event that contains all possible events. Sample space and events are primitive (undefined) notions (cf. line and point in geometry).

5.2.1 Generalized Union

The event that at least one of N events $A_1, A_2, \ldots, A_N$ occurs (i.e., $A = A_1 \cup A_2 \cup \cdots \cup A_N$) needs information not only about the individual events but about all possible overlaps.

Theorem. The probability P_1 of the realization of at least one among the events $A_1, A_2, \ldots, A_N$ is given by

$$P_1 = S_1 - S_2 + S_3 - S_4 + \cdots \pm S_N \,, \tag{5.8}$$

where the S_r are defined as the sums of all probabilities with r subscripts (e.g., $S_1 = \sum p_i$, $S_2 = \sum p_{ij}$, and $i < j < k < \cdots \leq N$) so that each contribution appears only once; hence, each sum S_r has $\binom{N}{r}$ terms, and the last term S_N gives the probability of the simultaneous realization of all terms.[6]

This result can be used to solve an old problem. Consider two sequences of N unique symbols differing only in the order of occurrence of the symbols and which are then compared, symbol by symbol. What is the probability P_1 that there is at least one match? Let A_k be the event that a match occurs at the kth position. Therefore, symbol number k is at the kth place, and the remaining $N - 1$ are anywhere; hence,

$$p_k = \frac{(N-1)!}{N!} = \frac{1}{N} \,,$$

and for every combination i, j,

$$p_{ij} = \frac{(N-2)!}{N!} = \frac{1}{N(N-1)} \,.$$

Each term in S_r in Eq. (5.8) equals $(N - r)!/N!$ and therefore $1/r!$; therefore,

$$P_1 = 1 - \frac{1}{2!} + \frac{1}{3!} - + \cdots \pm \frac{1}{N!} \,. \tag{5.9}$$

One might recognize that $1 - P_1$ represents the first $N + 1$ terms in the expansion of $1/e$; hence, $P_1 \approx 1 - 1/e \approx 0.632$. It seems rather remarkable that P_1 is independent of N. For problems of matching genes and the like it is useful to consider an extension, that for any integer $1 \leq m \leq N$ the probability $P_{[m]}$ that exactly m among the N events $A_1, \ldots, A_N$ occur simultaneously is[6]

$$P_{[m]} = S_m - \binom{m+1}{m} S_{m+1} + \binom{m+2}{m} S_{m+2} - + \cdots \pm \binom{N}{m} S_N \tag{5.10}$$

---

[6]The proof is given in Feller (1967), Chap. 4.

and

$$P_{[0]} = 1 - P_1 = 1 - 1 + \frac{1}{2!} - \frac{1}{3!} + - \cdots \pm \frac{1}{(N-2)!} \mp \frac{1}{(N-1)!} \pm \frac{1}{N!}$$

$$P_{[1]} = 1 - 1 + \frac{1}{2!} - \frac{1}{3!} + - \cdots \pm \frac{1}{(N-2)!} \mp \frac{1}{(N-1)!}$$

$$P_{[2]} = \frac{1}{2!}\left[1 - 1 + \frac{1}{2!} - \frac{1}{3!} + - \cdots \pm \frac{1}{(N-3)!} \mp \frac{1}{(N-2)!}\right]$$

$$\vdots$$

$$P_{[N-1]} = \frac{1}{(N-1)!}\{1-1\} = 0$$

$$P_{[N]} = \frac{1}{N!} .$$

Noticing again the similarity with the expansion of $1/e$, for large N,

$$P_{[m]} \approx \frac{e^{-1}}{m!} \tag{5.11}$$

(i.e., a special case of the Poisson distribution with $\lambda = 1$). The probability P_m that m or more of the events $A_1, \ldots, A_N$ occur simultaneously is

$$P_m = P_{[m]} + P_{[m+1]} + \cdots + P_{[N]}. \tag{5.12}$$

Starting with Eq. (5.9) and noting that

$$P_{[m+1]} = P_m - P_{[m]} , \tag{5.13}$$

by induction, for $m \geq 1$,

$$P_{[m]} = S_m - \binom{m}{m-1}S_{m+1} + \binom{m+1}{m-1}S_{m+2}$$
$$- \binom{m+2}{m-1}S_{m+3} + \cdots \pm \binom{N-1}{m-1}S_N . \tag{5.14}$$

5.2.2 Conditional Probability

The notion of *conditional probability* is of great importance.[7] It refers to questions of the type "what is the probability of event A, given that H has occurred?" We use the notation $P\{A|H\}$ (read as "the conditional probability of A on hypothesis H" or "the conditional probability of A for a given event H") and

$$P\{A|H\} = \frac{P\{AH\}}{P\{H\}} . \tag{5.15}$$

---

[7]Indeed, Reichenbach, Popper, and others have taken the view that conditional probability may and should be chosen as the basic concept of probability theory. We should in any case note that most of the results derived for unconditional probabilities are also valid for conditional probabilities.

This result can be derived by noting that we are asking "to what extent is H contained in A?" which means "to what extent are H and A likely to occur simultaneously?" In set notation, this is $P\{A \cap H\} = P\{H \cap A\}$. Therefore, $P\{A|H\} = kP\{A \cap H\}$, where k is a constant. If $A = H$, then $P\{H|H\} = kP\{H \cap H\} = kP\{H\} = 1$; hence, $k = 1/P\{H\}$ and we obtain

$$P\{A|H\} = \frac{P\{A \cap H\}}{P\{H\}} \tag{5.16}$$

(i.e., Eq. 5.15). If all sample points have equal probabilities, then

$$P\{A|H\} = \frac{N\{AH\}}{N\{H\}} \ , \tag{5.17}$$

where $N\{AH\}$ is the number of sample points common to A and H.

From this comes a theorem, due to Bayes, of great importance and widely referred to, which gives the probability that the event A, which has occurred, is the result of the cause E_k:

$$P\{E_k|A\} = \frac{P\{A|E_k\}P\{E_k\}}{\sum_{j=1}^{n} P\{A|E_j\}P\{E_j\}} \quad \text{for } k = 1, \dots, n \ , \tag{5.18}$$

where the E_j are mutually exclusive hypotheses.

Proof. Let the simple events E_i be labelled such that

$$A = E_1 \cup E_2 \cup \cdots \cup E_m \ , \quad 1 \leq m \leq n \ . \tag{5.19}$$

Then

$$P\{A\} = \sum_{j=1}^{m} P\{E_j\} \ . \tag{5.20}$$

From the definition (5.15),

$$\sum_{j=1}^{n} P\{A|E_j\}P\{E_j\} = \sum_{j=1}^{n} P\{A \cap E_j\} \ , \tag{5.21}$$

which can be equated to the right-hand side of (5.20)

$$\sum_{j=1}^{n} P\{A \cap E_j\} = \sum_{j=1}^{m} P\{E_j\} = P\{A\} \ . \tag{5.22}$$

This result can be used to write the denominator of the right-hand side of Eq. (5.18) as $P\{A|E_k\}P\{E_k\}/P\{A\}$, but this, according to Eq. (5.16) and after cancelling equals $P\{A \cap E_k\}/P\{A\} = P\{E_k \cap A\}/P\{A\}$, which, again using Eq. (5.16), equals $P\{E_k|A\}$. $\qquad\qquad\qquad\qquad\qquad\qquad\qquad\qquad\qquad\qquad\qquad\qquad\qquad\qquad\quad \square$

5.2.3 Bernoulli Trials

Bernoulli trials are defined as repeated, (stochastically) independent trials[8] (hence, probabilities multiply) with only two possible outcomes per trial—success (s) or failure (f)—with respective constant (throughout the sequence of trials) probabilities p and $q = 1 - p$. The sample space of each trial is $\{s, f\}$, and the sample space of n trials contains 2^n points. The event "k successes, with $k = 0, 1, \ldots, n$, and $n - k$ failures in n trials" can occur in as many ways as k letters can be distributed among n places (the order of successes and failures does not matter), and each of the $^nC_k = \binom{n}{k}$ points has probability $p^k q^{n-k}$. Hence, the probability of exactly k successes in n trials is

$$b(k; n, p) = \binom{n}{k} p^k q^{n-k} . \tag{5.24}$$

This function is known as the binomial distribution because the terms are those of the expansion of $(a + b)^n$ (cf. Sect. 4.3).

Bernoulli trials are easily generalized to more than two outcomes. If the probability of realizing an outcome E_i is p_i ($i = 1, 2, \ldots, r$) subject only to the condition

$$p_1 + p_2 + \cdots + p_r = 1 , \tag{5.25}$$

then the probability that in n trials, E_1 occurs k_1 times, E_2 occurs k_2 times, and so on is

$$\frac{n!}{k_1! k_2! \cdots k_r!} p_1^{k_1} p_2^{k_2} \cdots p_r^{k_r} , \tag{5.26}$$

where

$$k_1 + k_2 + \cdots + k_r = n . \tag{5.27}$$

The reader can readily verify that a plot of b versus k is a hump whose central term occurs at $m = [(n + 1)p]$, where the notation $[x]$ signifies "the largest integer not exceeding x".

An important practical case arises where n is large and p is small, such that the product $np = \lambda$ is of moderate size (~ 1). The distribution can then be simplified:

$$b(k; n, p) = \binom{n}{k} \frac{\lambda^k}{n} \left(1 - \frac{\lambda}{n}\right)^{n-k} = \frac{\lambda^k}{k!} \left(1 - \frac{\lambda}{n}\right)^{n-k} \frac{n(n-1)\cdots(n-k+1)}{n^k} .$$

Now, $(1 - \lambda/n)^{n-k} \approx e^{-\lambda}$ and $n(n-1)\ldots(n-k+1)/n^k \approx 1$; hence,

$$b(k; n, p) \approx \frac{\lambda^k}{k!} e^{-\lambda} = p(k; \lambda) , \tag{5.28}$$

---

[8]Stochastic independence is formally defined via the condition

$$P\{AH\} = P\{A\}P\{H\} , \tag{5.23}$$

which must hold if the two events A and H are stochastically (sometimes called statistically) independent.

which is called the Poisson approximation to the binomial distribution. However, if λ is fixed, then $\sum p(k; \lambda) = 1$; hence, $p(k; \lambda)$, the probability of exactly k successes occurring, is a distribution in its own right, called the Poisson distribution. It is of great importance in nature, describing processes lacking memory.

The probability $f(k; r, p)$ that exactly k failures precede the rth success (i.e., exactly k failures among $r + k - 1$ trials followed by success) is

$$f(k; r, p) = \binom{r + k - 1}{k} p^r q^k = \binom{-r}{k} p^r (-q)^k, \quad k = 0, 1, 2, \ldots . \quad (5.29)$$

Iff[9]

$$\sum_{k=0}^{\infty} f(k; r, p) = 1 , \quad (5.30)$$

the possibility that an infinite sequence of trials produces fewer than r successes can be discounted, since by the binomial theorem

$$\sum_{k=0}^{\infty} \binom{-r}{k} (-q)^k = p^{-r} , \quad (5.31)$$

which equals 1 when multiplied by p^r. The sequence $f(k; r, p)$ is called the negative binomial distribution.

Example. Suppose that the normal rate of infection of a certain disease in cattle is 25 %.[10] An experimental vaccine is injected into n animals. If it is wholly ineffectual, the probability that exactly k animals remain free from infection is $b(k; n, 0.75)$; for $k = n = 10$, this probability is approximately 0.056; the probability that 1 animal out of 17 becomes infected is slightly lower, approximately 0.050, and for 2 out of 23, it is lower still, approximately 0.049. This example highlights the difficulties of drawing inferences from small samples. Two failures out of 23 is slightly better evidence in favour of the vaccine than no failures out of 10.

5.3   Moments of Distributions

A *random variable* is "a function defined on a sample space" (e.g., the number of successes in n Bernoulli trials). A unique rule associates a number $\mathbf{X}$ with any sample point. The aggregate of all sample points on which $\mathbf{X}$ assumes the fixed value x_j forms the event that $\mathbf{X} = x_j$, with probability $P\{\mathbf{X} = x_j\}$.[11] The function $f(x_j) = P\{\mathbf{X} = x_j\}$ is called the (probability) distribution of the random

---

[9]If and only if.
[10]Due to P.V. Sukhatme and V.G. Panse, quoted by Feller (1967), Chap. 6.
[11]$\mathbf{X}$ may assume the values $x_1, x_2, \ldots$ (i.e., the range of $\mathbf{X}$).

variable $\mathbf{X}$.[12] Joint distributions are defined for two or more variables defined on the same sample space. For two variables, $p(x_j, y_k) = P\{\mathbf{X} = x_j, \mathbf{Y} = y_k\}$ is the joint probability distribution of $\mathbf{X}$ and $\mathbf{Y}$.

The mean, average, or expected value of $\mathbf{X}$ is defined by[13]

$$\mu_X = \mathbf{E}(\mathbf{X}) = \sum x_k f(x_k) \tag{5.33}$$

provided that the series converges absolutely. The expectation of the sum (or product) of random variables is the sum (or product) of their expectations. Proofs are left to the reader.

Any function of $\mathbf{X}$ may be substituted for $\mathbf{X}$ in definition (5.33), with the same proviso of series convergence. The expectations of the rth powers of $\mathbf{X}$ are called the rth moments of $\mathbf{X}$ about the origin.[14] Since $|\mathbf{X}|^{r-1} \leq |\mathbf{X}|^r + 1$, if the rth moment exists, so do all the preceding ones. The expectation of the square of $\mathbf{X}$'s deviation from its mean value has a special name, the variance:[15]

$$\sigma_X^2 = \mathrm{Var}(\mathbf{X}) = \mathbf{E}((\mathbf{X} - \mathbf{E}(\mathbf{X}))^2) = \mathbf{E}(\mathbf{X}^2) - \mathbf{E}(\mathbf{X})^2 . \tag{5.34}$$

Its positive square root σ is called the standard deviation of $\mathbf{X}$, hinting at its use as a rough measure of spread. The mean and variance (i.e., the first and second moments) provide a convenient way to normalize (render dimensionless) a random variable, namely

$$\mathbf{X}^* = \frac{\mathbf{X} - \mu_X}{\sigma_X} . \tag{5.35}$$

The covariance measures the linear association between variables $\mathbf{X}$ and $\mathbf{Y}$ and is defined as

$$\mathrm{Cov}(\mathbf{X}, \mathbf{Y}) = \mathbf{E}(\mathbf{X} - \mathbf{E}(\mathbf{X}))\mathbf{E}(\mathbf{Y} - \mathbf{E}(\mathbf{Y})) = \mathbf{E}(\mathbf{X}\mathbf{Y}) - \mathbf{E}(\mathbf{X})\mathbf{E}(\mathbf{Y}) \tag{5.36}$$

(explicity, as $(1/n) \sum_{j=1}^n (x_j - \mu_X)(y_j - \mu_Y)$). It equals zero if the variables are uncorrelated. If more than two variables are involved, it is convenient to arrange the pairwise covariances in the so-called covariance matrix.

The scatter matrix S of n samples of m-dimensional data is defined as

$$S = \sum_{j=1}^n (\mathbf{X}_j - \mathbf{E}(\mathbf{X}))(\mathbf{X}_j - \mathbf{E}(\mathbf{X}))^{\mathrm{T}} . \tag{5.37}$$

---

[12]The distribution function $F(x)$ of $\mathbf{X}$ is defined by

$$F(x) = P\{\mathbf{X} \leq x\} = \sum_{x_j \leq x} f(x_j) \tag{5.32}$$

(i.e., a nondecreasing function tending to 1 as $x \to \infty$).

[13]Also denoted by angular brackets or a bar.

[14]Notice the mechanical analogies: centre of gravity as the mean of a mass and moment of inertia as its variance.

[15]Older literature uses the term "dispersion".

If the variables are normally distributed, the (normalized) scatter matrix provides an estimate of the covariance matrix.

Problem. Calculate the means and variances of the binomial and Poisson distributions.

5.3.1 Runs

Studies of the statistical properties of DNA and the like often start by stating the total numbers of the four bases A, C, T, and G. This information entirely neglects information on the order in which they occur. The theory of the distribution of runs is one way of handling this information. A run is defined as a succession of similar events preceded and succeeded by different events; the number of elements in a run will be referred to as its length. The number of runs of course equals the number of unlike neighbours.

Here, we shall only derive the distribution of runs of two kinds of elements. More complicated results may be found by reference to Mood's (1940) paper.

Let the two kinds of elements be a and b (they could be purines and pyrimidines), and let there be n_1 as and n_2 bs, with $n_1 + n_2 = n$. r_{1i} will denote the number of runs of a of length i, with $\sum_i r_{1i} = r_1$, and so on. It follows that $\sum i r_{1i} = n_1$, and so on. Given a set of as and bs, the numbers of different arrangements of the runs of a and b are given by multinomial coefficients and the total number of ways of obtaining the set r_{ji} ($j = 1, 2; i = 1, 2, \ldots, n_1$) is

$$N(r_{ji}) = \begin{bmatrix} r_1 \\ r_{1i} \end{bmatrix} \begin{bmatrix} r_2 \\ r_{2i} \end{bmatrix} F(r_1, r_2) , \tag{5.38}$$

where the special function $F(r_1, r_2)$ is the number of ways of arranging r_1 objects of one kind and r_2 objects of another so that no two adjacent objects are of the same kind (see Table 5.1).

Since there are $\binom{n}{n_1}$ possible arrangements of the as and bs, the distribution of the r_{ji} is

$$P(r_{ji}) = \frac{N(r_{ji}) F(r_1, r_2)}{\binom{n}{n_1}} . \tag{5.39}$$

Table 5.1 Values of the function $F(r_1, r_2)$

$r_1 - r_2$	$F(r_1, r_2)$
> 1	0
1	1
0	2

5.3.2 The Hypergeometric Distribution

Continuing the notation of the previous subsection, consider choosing r elements at random from the binary mixture of as and bs. What is the probability q_k that the group will contain exactly k as? It must necessarily contain $r - k$ bs, and the two types of elements can be chosen in $\binom{n_1}{k}$ and $\binom{n-n_1}{r-k}$ ways, respectively. Since any choice of k as can be combined with any choice of $r - k$ bs,

$$q_k = \frac{\binom{n_1}{k}\binom{n-n_1}{r-k}}{\binom{n}{r}} . \tag{5.40}$$

This system of probabilities is called the hypergeometric distribution (because the generating function of q_k is expressible in terms of hypergeometric functions). Many combinatorial problems can be reduced to this form.

Problem. A protein consists of 300 amino acids, of which it is known that there are 2 cysteines. A 50-mer fragment has been prepared. What are the probabilities that 0, 1, or 2 cysteines are present in the fragment?

5.3.3 Multiplicative Processes

Many natural processes are *random additive processes*; for example, a displacement is the sum of random steps (to the left or to the right in the case of the one-dimensional random walk; cf. Chap. 6). The probability distribution of the net displacement after n steps is the binomial function. The central limit theorem guarantees that this distribution is Gaussian as $n \to \infty$, a universal property of random additive processes.

Although their formalism is less familiar, *random multiplicative processes* (RMP) are not less common in nature. An example is rock fragmentation. From an initial value x_0, the size of a rock undergoing fragmentation evolves as $x_0 \to x_1 \to x_2 \to \cdots \to x_N$. If the size reduction factor

$$r_n = \frac{x_n}{x_{n-1}} \tag{5.41}$$

is less than 1, we have

$$x_N = x_0 \prod_{k=1}^{N} r_k . \tag{5.42}$$

Extreme events, although exponentially rare, are exponentially different. Hence, *the average is dominated by rare events*. This is quite different from the more intuitively acceptable random additive process. If the phenomenon is of that type, the more measurements one can take, the better the estimate of its value. However, if the phenomenon is an RMP, as one increases the number of measurements, the estimate of the mean will fluctuate more and more, before ultimately converging to a stable value. Since multiplication is equivalent to adding logarithms, it is not surprising that the distribution of the result of an RMP is lognormal (i.e., $\ln p = \sum \ln p_i$), and the average value (expectation) of p is

$$\bar{p} = \sum_{n=0}^{N}(Nn)p^n q^{N-n} \ . \tag{5.43}$$

5.4   Likelihood

The search for *regularities* in nature has already been mentioned as the goal of scientific work. Often, these regularities are framed in terms of *hypotheses*.[16] With hypotheses (which may eventually become theories), laws and relations acquire more than immediate validity and relevance (cf. unconditional information, Sect. 2.1.1).

In observing the natural world, one encounters "deterministic" events, character-ized by rather clear relationships between the quantities measured compared with the experimental uncertainties, and more uncertain events with statistical outcomes (such as coin tossing or Mendelian gene segregation). The latter raise the general problem of how to assess the relative merits of alternative hypotheses in the light of the observed data. Statistics concerns itself with tests of significance and with estimation (i.e., seeking acceptable values for the parameters of the distributions specified by the hypotheses).

The *method of support* proposes that

posterior support = prior support + experimental support

and

$$\text{information gained} = \log \frac{\text{posterior probability}}{\text{prior probability}} \ .$$

Two rival approaches to estimation have arisen: the theory of inverse probabil-ity (due to Laplace), in which the probabilities of causes (i.e., the hypotheses) are deduced from the frequencies of events, and the method of likelihood (due to Fisher). In the theory of inverse probability, these probabilities are interpreted as quantitative and absolute measures of belief. Although it still has its adherents, the system of inference based on inverse probability suffers from the weakness of supposing that hypotheses are selected from a continuum of infinitely many hypotheses. The prior probabilities have to be invented; for example, by imagining a chance setup, in which case the model is a private one and violates the principle of public demonstrability. Alternatively, one can apply Laplace's "Principle of Insufficient Reason," according

---

[16]Strictly speaking, one should instead refer to propositions. A hypothesis is an asserted proposition, whereas at the beginning of an investigation it would be better to start with considered propositions, to avoid prematurely asserting what one wishes to find out. Unfortunately, the use of the term "hypothesis" seems to have become so well established that we may risk confusion if we avoid using the word.

to which each hypothesis is given the same probability if there are no grounds to believe otherwise. Conceptually, that viewpoint is rather hard to accept. Moreover, if there are infinitely many equiprobable hypotheses, then each one has an infinitesimal probability of being correct.

Bayes' theorem (5.18) may be applied to the weighting of hypotheses if and only if the model adopted includes a chance setup for the generation of hypotheses with specific prior probabilities. Without that, the method becomes one of inverse probability. Equation (5.18) is interpreted as equating the posterior probability of the hypothesis E_k (after having acquired data A) to our prior estimate of the correctness of E_k (i.e., before any data were acquired), $P\{E_k\}$, multiplied by the prior probability of obtaining the data given the hypothesis (i.e., the likelihood; see below), the product being normalized by dividing by the sum over all hypotheses.

A fundamental critique of Bayesian methods is that the Bayes–Laplace approach regards hypotheses as being drawn at random from a population of hypotheses, a certain proportion of which is true. "Bayesians" regard it as a strength that they can include prior knowledge, or rather prior states of belief, in the estimation of the correctness of a model. Since that appears to introduce a wildly fluctuating subjectivity into the calculations, it seems more reasonable to regard that as a fatal weakness of the method.[17]

To reiterate: our purpose is to find what is the most likely explanation of a set of observations; that is, a description that is simpler, hence shorter, than the set of facts observed to have occurred.[18]

The three pillars of statistical inference are as follows:

1. A statistical model: that part of the description that is not (at least at present) in question (corresponding to K in Eq. 2.13).
2. The data: that which has been observed or measured (unconditional information);
3. The statistical hypothesis: the attribution of particular values to the unknown parameters of the model that are under investigation (conditional information).

The preferred values of those parameters are then those that maximize the likelihood of the model, likelihood being defined in the following:

Definition. The likelihood $L(H|R)$ of the hypothesis H given data R and a specific model is proportional to $P(R|H)$, the constant of proportionality being arbitrary but constant in any one application (i.e., with the same model and the same data, but different hypotheses).

The arbitrariness of the constant of proportion is of no concern since, in practice, likelihood ratios are taken, as in the following:

---

[17]As Fisher and others have pointed out, it is not strictly correct to associate Bayes with the inverse probability method. Bayes' doubts as to its validity led him to withhold publication of his work (it was published posthumously).

[18]Sometimes brevity is taken as the main criterion. This is the minimum description length (MDL) approach. See also the discussion in Sects. 3.4 and 6.5.

Definition. The likelihood ratio of two hypotheses on some data is the ratio of their likelihoods on that data. It will be denoted as $L(H_1, H_2|R)$. The likelihood ratio of two hypotheses on independent sets of data may be multiplied together to form the likelihood ratio on the combined data:

$$L(H_1, H_2|R_1 \& R_2) = L(H_1, H_2|R_1) \times L(H_1, H_2|R_2) . \tag{5.44}$$

The fundamental difference between probability and likelihood is that in the inverse probability approach R is variable and H constant, whereas in likelihood, H is variable and R constant. In other words, likelihood is predicated on a fixed R.

We shall sometimes need to recall that if R_1 and R_2 are two possible, mutually exclusive, results and $P\{R|H\}$ is the probability of obtaining the result R given H, then

$$P\{R_1 \text{ or } R_2|H\} = P\{R_1|H\} + P\{R_2|H\} \tag{5.45}$$

and

$$P\{R_1 \text{ and } R_2|H\} = P\{R_1|H\}P\{R_2|H\} . \tag{5.46}$$

The method of likelihood reposes on the definitions of likelihood *per se* and of the likelihood ratio.

Example. The problem is to determine the probability that a baby will be a boy. We take a binomial model (cf. Sect. 5.2.3) for the occurrence of boys and girls in a family of two children; we have two sets of data—R_1: one boy and one girl, and R_2: two boys—and two hypotheses—H_1: the probability p of a birth being male born equals $\frac{1}{4}$, and H_2: $p = \frac{1}{2}$. Hence,

$P\{R\|H\}$	R_1	R_2
H_1	$2p(1-p) = \frac{3}{8}$	$p^2 = \frac{1}{16}$
H_2	$2p(1-p) = \frac{1}{2}$	$p^2 = \frac{1}{4}$

By inspection, $P\{R|H\}$ for H_2 exceeds that for H_1 for both sets of data, from which we may infer that H_2 is better supported by the data.

The concept of likelihood ratio can easily be extended to continuous distributions; that is, $P\{R|H\}$ becomes a probability density. The likelihood ratio is computed for the distribution with respect to one value chosen arbitrarily and the maximum is sought. Usually it is better to work in logarithms, and the *support* $\mathfrak{S}$ is defined as the logarithm of the likelihood, namely

$$\mathfrak{S}(p) = \log L(p) . \tag{5.47}$$

The curvature of $\mathfrak{S}(p)$ at its maximum has been called the information, and its reciprocal is a natural measure of the uncertainty about p (i.e., the width of the peak is inversely related to the degree of certainty of the estimation).

The method of maximum likelihood provides the ability to deliver a conclusion compatible with the given evidence.

5.5   The Maximum Entropy Method

Consider the problem of deducing the positions of stars and galaxies from a noisy map of electromagnetic radiation intensity. One should have an estimate for the average noise level: The simple treatment of such a map is to reject every feature greater than the mean noise level and accept every one that is greater. Such a map is likely to be a considerably distorted version of reality.[19]

The maximum entropy method can be considered as a heuristic drill for applying D. Bernoulli's maxim: "Of all the innumerable ways of dealing with errors of observation, one should choose the one which has the highest degree of probability for the complex of observations as a whole" (cf. footnote 18 in Chap. 2). In effect, it is a generalization of the method of maximum likelihood.

First, the experimental map must be digitized both spatially and with respect to intensity; that is, it is encoded as a finite set of pixels, each of which may assume one of a finite number of density levels. Let that density be m_j at the jth pixel. Then random maps are generated and compared with the data. All those inconsistent with the data (with due regard to the observational errors) are rejected. The commonest map remaining is then the most likely representation. This process is the constrained maximization of the configurational entropy $-\sum m_j \log m_j$ (the unconstrained maximization would simply lead to a uniform distribution of density over the pixels). Maximum entropy image restoration yields maximum information in Shannon's sense.

References

Feller W (1967) An introduction to probability theory and its applications, vol 1, 3rd edn. Wiley, New York
von Mises R (1931) Wahrscheinlichkeitsrechung. Deuticke, Leipzig
Mood AM (1940) The distribution theory of runs. Ann Math Stat 11:367–392
Planck M (1932) The concept of causality. Proc Phys Soc 44:529–539
Sommerhoff G (1950) Analytical biology. Oxford University press, London

---

[19]Implicitly, Platonic reality is meant here.

Randomness and Complexity

6

Randomness is a concept deeply entangled with bioinformatics. A random sequence cannot convey information, in the sense that it could be generated by a recipient merely by tossing a coin. Randomness is therefore a kind of "null hypothesis"; a random sequence of symbols is a sequence lacking all constraints limiting the variety of choice of successive symbols selected from a pool with constant composition (i.e., an ergodic source). Such a sequence has maximum entropy in the Shannon sense; that is, it has minimum redundancy.

If we are using such an ideally random sequence as a starting point for assessing departures from randomness, it is important to be able to recognize this ideal randomness. How easy is this task? Consider the following three sequences:

$$11111111111111111111111111111111$$
$$01010101010101010101010101010101$$
$$10010100010100101010111101010101010$$

each of which could have been generated by tossing a coin. According to the results from the previous two chapters, all three outcomes, indeed any sequence of 32 1s and 0s, have equal probability of occurrence, namely $1/2^{32}$. Why do the first two not "look" random? Kolmogorov supposed that the answer might belong to psychology; Borel even asserted that the human mind is unable to simulate randomness (presumably the ability to recognize patterns was—and is—important for our survival). Yet, apparent pattern is also present in random sequences: van der Waerden (1927) has proved that in every infinite binary sequence at least one of the two symbols must occur in arithmetical progressions of every length. Hence, the first of the above three sequences would be an unexceptionable occurrence in a much longer random sequence—in fact, whether a given sequence is random is formally undecidable. At best, then, we can hope for heuristic clues to the possible absence of randomness and, hence, presumably the presence of meaning, in a gene sequence.

In anticipation of the following sections, we can already note that incompressibility (i.e., the total absence of regularities) forms a criterion of randomness. This criterion uses the notion of algorithmic complexity. The first sequence can be generated by the brief instruction "write '1' 32 times" and the second by the only marginally

© Springer-Verlag London 2015
J. Ramsden, *Bioinformatics*, Computational Biology 21,
DOI 10.1007/978-1-4471-6702-0_6

longer statement "write '01' 16 times," whereas the third, which was generated by blindly tapping on a keyboard, has no apparent regularity.

"Absence of pattern" corresponds to the dictionary synonym "haphazard" (cf. the French expression "au hasard"). By counting the number of 1s and 0s in a long segment of the third sequence, we can obtain an estimate of the probability of occurrence of each symbol. "Haphazard" then means that the choice of each successive symbol is made independently, without reference to the preceding symbol or symbols, in sharp contrast to the second sequence, which could also be generated by the algorithm "if the preceding symbol is 1, write 0, otherwise write 1" operating on a starting seed of 1 or 6.

Note how closely this exercise of algorithmic compression is related to the general aim of science: to find the simplest set of axioms that will enable all the observable phenomena studied by the branch of science concerned to be explained (an empirical fact being "explained" if the propositions expressing it can be shown to be a consequence of the axioms constituting the scientific theory underpinning that branch). For example, Maxwell's equations turned out to be suitable for explaining the phenomena of electromagnetism.[1]

The meaning of randomness as denoting independence from what has gone before is well captured in the familiar expression "random access memory," the significance being that a memory location can be selected arbitrarily (cf. the German "beliebig", at whim), as opposed to a sequential access memory, whose elements can only be accessed one after the other. Mention of memory brings to mind the fact that successive independent choices implies the absence of memory in the process generating those choices.

The validity of the above is independent of the actual probabilities of choosing symbols; that is, they may be equal or unequal. Although in many organisms it turns out that the frequencies of occurrence of all four bases are in fact equal, this is by no means universal, it being well known that thermophilic bacteria have more C≡G base pairs than A=T in their genes, since the former, being linked by three hydrogen bonds, are more thermally stable than the latter, which only have two (cf. Fig. 11.3). Yet, we can still speak of randomness in this case. In binary terms, it corresponds to unequal probabilities of heads or tails, and the sequence may still be algorithmically incompressible; that is, it cannot be recreated by any means shorter than the process actually used to generate it in the first place.

---

[1] An obvious corollary of this association of randomness with algorithmic compressibility is that there is an intrinsic absurdity in the notion of an algorithm for generating random numbers, such as those included with many compilers and other software packages. These computer-generated pseudorandom numbers generally pass the usual statistical tests for randomness, but little is known about how their nonrandomness affects results obtained using them. Quite possibly the best heuristic sources of (pseudo)random digits are the successive digits of irrational numbers like π or $\sqrt{2}$. These can be generated by a deterministic algorithm and, of course, are always the same, but in the sense that one cannot jump to (say) the hundredth digit without computing those preceding it, they do fulfil the criteria of haphazardness.

We have previously stated that bioinformatics could be considered to be the study of the departures from randomness of DNA. We are shown a sequence of DNA: Is it random? We want to be able to quantify its departure from randomness. Presumably those sequences belonging to viable organisms, or even to their individual proteins or promoter sequences, are not random. What about introns, and intergenome sequences? If they are indeed "junk," as is sometimes (facetiously?) asserted, then we might well expect them to be random. Even if they started their existence as nonrandom sequences, they may have been randomized since they would be subject to virtually no selection pressure. Mutations are supposed to be random and occur at random places. The opposite procedure would be that all DNA sequences started as random ones and then natural selection eliminated many according to some systematic criterion; therefore, the extant collection of the DNA of viable organisms on this planet is not random. Can we, then, say anything about the randomness or otherwise of an individual sequence taken in isolation?

Similar considerations apply to proteins. Given a collection of amino acid sequences of proteins (which, to be meaningful, should come from the same genome), we can assess the likelihood that they arose by chance and the degree of their departures from randomness.

All such sequences can be idealized as sequences of Bernoulli trials (see Sect. 5.2.3), which are themselves abstractions of a coin tossing experiment. Since order does not matter in determining the probability of a given overall outcome, 50 heads followed by 50 tails has the same probability of occurring as 50 alternations of heads and tails, which again is no less probable than a particular realization in which the heads and tails are "randomly" mixed.

Any nonbinary sequence can, of course, be encoded in binary form. Typical procedures for biological sequences (amino acids or nucleotides) are to consider nucleotides as purines (0) or pyrimidines (1), or amino acids as hydrophobic (apolar) or hydrophilic (polar) residues (cf. Markov's encoding of poetry as a sequence of vowels and consonants). Alternatively, the nucleotides could constitute a sequence in base 4 ($A \equiv 0$, $C \equiv 1$, $T \equiv 2$, $G \equiv 3$), which can then be converted to base 2.

It is a commonly held belief that after a long sequence of heads (say), the opposite result (tails) becomes more probable. There is no empirical support for this assertion in the case of coin tossing. In other situations in which the outcome depends on selecting elements from a finite reservoir, however, clearly this result must hold. Thus, if a piece of DNA is being assembled from a soup of base monomers at initially equal concentrations, if by chance the sequence starts out by being poor in A, say, then later on this must be compensated by enrichment (chain elongation ends when all available nucleotides have been consumed).

Formal Notions of Randomness In order to proceed further, we need to more carefully understand what we mean by randomness. Despite the fact that the man in the street supposes that he has a good idea of what it means, randomness is a rather delicate concept. The toss of an unbiased coin is said to be random; the probability of heads or tails is 0.5. We cannot assess the randomness of a single result, but we can assess the probability that a sequence of tosses is random. So perhaps we can answer the question of whether a given individual sequence is random. The three main notions of randomness are as follows:[2]

1. Stochasticity, or frequency stability, associated with von Mises, Wald, and Church;[3]
2. Incompressibility or chaoticity, associated with Solomonoff, Kolmogorov, and Chaitin;[4]
3. Typicality, associated with Martin-Löf (and essentially coincident with incompressibility).

6.1   Random Processes

A process characterized by a succession of values of a characteristic parameter y is called random if y does not depend in a completely definite way on the independent variable, usually (laboratory) time t, but in the context of sequences, the independent variable could be the position along the sequence. A random process is therefore essentially different from a causal process (cf. Sect. 5.1). It can be completely defined by the set of probability distributions $W_1(yt)dy$, the probability of finding y in the range $(y, y + dy)$ at time t, $W_2(y_1t_1, y_2t_2)\,dy_1\,dy_2$, the joint probability of finding y in the range $(y_1, y_1 + dy_1)$ at time t_1 and in the range $(y_2, y_2 + dy_2)$ at time t_2, and so forth for triplets, quadruplets, ... of values of y.

If there is an unchanging underlying mechanism, the probabilities are stationary and the distributions can be simplified as $W_1(y)dy$, the probability of finding y in the range $(y, y + dy)$; $W_2(y_1y_2t)\,dy_1\,dy_2$, the joint probability of finding y in the ranges $(y_1, y_1 + dy_1)$ and $(y_2, y_2 + dy_2)$ separated by an interval of time $t = t_2 - t_1$; and so on. Experimentally, a single long record $y(t)$ can be cut into pieces (which should be longer than the longest period supposed to exist), rather than carrying out measurements on many similarly prepared systems. This equivalence of time

---

[2] After Volchan (2002).

[3] von Mises called the random sequences in accord with this notion "collectives". It was subsequently shown that the collectives were not random enough (see Volchan (2002) for more details); for example, the number 0.0123456789101112131415161718192021 ... satisfied von Mises' criteria but is clearly computable.

[4] The Kolmogorov-Chaitin definition of the descriptive or algorithmic complexity $K(s)$ of a symbolic sequence s with respect to a machine M running a program P is given by

$$K(s) = \begin{cases} \infty & \text{if there is no P such that } M(\mathrm{P}) = s \\ \min\{|\mathrm{P}| : M(\mathrm{P}) = s\} & \text{otherwise .} \end{cases} \qquad (6.1)$$

This means that $K(s)$ is the size of the smallest input program P that prints s and then stops when input into M. In other words, it is the length of the shortest (binary) program that describes (codifies) s. Insofar as M is usually taken to be a universal Turing machine, the definition is machine-independent.

and ensemble averages is called ergodicity. Note, however, that many biological systems appear to be frozen in small regions of state space, as a glass, and hence are nonergodic (cf. Sect. 3.4.2).

Notice some of the difficulties inherent in the above description. For example, we referred to "an unchanging underlying mechanism," yet at the same time asserted that a random process is one which does not depend in a completely definite way on the independent variable. Yet, who would deny that the coin, whose tossing generates that most archetypical of random sequences, does not follow Newton's laws of motion? This apparent paradox can be shown to be a consequence of dynamic chaos (Sect. 7.3).

If successive values of y are not correlated at all, that is,

$$W_2(y_1 t_1, y_2 t_2) = W_1(y_1 t_1) W_1(y_2 t_2) \tag{6.2}$$

etc., all information about the process is completely contained in W_1 and the process is called a purely random process.

6.2 Markov Chains

In the previous section we considered "purely random" processes in which successive values of a variable, y, are not correlated at all. If, however, the next step of a process depends on its current state; that is,

$$W_2(y_1 y_2 t) = W_1(y_1) P_2(y_2 | y_1 t) , \tag{6.3}$$

where $P_2(y_2 | y_1 t)$ denotes the conditional probability that y is in the range $(y_2, y_2 + dy_2)$ after having been at y_1 at a time t earlier, we have a Markov chain.

Definition. A sequence of trials with possible outcomes **a** (possible states of the system), an initial probability distribution $\mathbf{a}^{(0)}$, and (stationary) transition probabilities defined by a stochastic matrix P is called a Markov chain.[5]

The probability distribution for an r-step process is

$$\mathbf{a}^{(r)} = \mathbf{a}^{(0)} P^r . \tag{6.4}$$

If the first m steps of a Markov process lead from a_j to some intermediate state a_i, then the probability of the subsequent passage from a_i to a_k does not depend on the manner in which a_i was reached, that is,

$$p_{jk}^{(m+n)} = \sum_i p_{ji}^{(m)} p_{ik}^{(n)} , \tag{6.5}$$

where $p_{jk}^{(n)}$ is the probability of a transition from a_j to a_k in exactly n steps (this is a special case of the Chapman–Kolmogorov identity).

---

[5]In some of the literature, one finds stochastic matrices arranged such that the columns rather than the rows sum to unity. The arrow in the top left-hand corner serves to indicate which convention is being used.

If upon repeated application of P the distribution **a** tends to an unchanging limit (i.e., an equilibrium set of states) that does not depend on the initial state, the Markov chain is said to be ergodic, and we can write

$$\lim_{r \to \infty} P^r = Q , \qquad (6.6)$$

where Q is a matrix with identical rows.[6] Now,

$$P P^n = P^n P = P^{n+1} , \qquad (6.7)$$

and if Q exists it follows, by letting $n \to \infty$, that

$$PQ = QP = Q \qquad (6.8)$$

from which Q (giving the stationary probabilities; i.e., the equilibrium distribution of **a**) can be found.

If all the transitions of a Markov chain are equally probable, then there is a complete absence of constraint; the process is purely random (a zeroth-order chain). Higher-order Markov processes have already been discussed (see Sect. 2.2).

A Markov chain represents an automaton (cf. Sect. 7.1.1) working incessantly. If the transformations were determinate (i.e., all entries in the transition matrix were 0 or 1), then the automaton would reach an attractor after a finite number of steps. The nondeterminate transformation can, however, continue indefinitely (although if any diagonal element is unity, it will get stuck there). If chains are nested inside one another, one has a hidden Markov model (HMM, see Sect. 13.5.2): suppose that the transformations accomplished by an automaton are controlled by a parameter that can take values a_1 or a_2, say. If a_1 is input, the automaton follows one matrix of transitions and if a_2 is input, it follows another set. The HMM is created if transitions between a_1 and a_2 are also Markovian. Markov chain Monte Carlo (MCMC) is used when the number of unknowns is itself an unknown.

One of the difficulties in the use of Markov chains to model processes is to ensure adequate statistical justification for any conclusions. The problem essentially concerns the inferences about the transition probabilities that one would like to make from a long, unbroken observation.[7] The problem becomes particularly acute when evidence for higher-order Markov chains is sought, when the quantity of data required might be unattainable. An important result is Whittle's formula giving the distribution of the transition count:

$$N_{uv}^{(n)}(F) = F_{uv}^* , \qquad (6.9)$$

where $N_{uv}^{(n)}(F)$ is the number of sequences $(a_1, a_2, \ldots, a_{n+1})$ having transition count $F = \{f_{ij}\}$, and satisfying $a_1 = u$ and $a_{n+1} = v$. The transition count together with the initial state (with probability p_{a_1}) forms a sufficient statistic for the process, since

$$p_{a_1} p_{a_1 a_2} \cdots p_{a_n a_{n+1}} = p_{a_1} \prod_{ij} p_{ij}^{f_{ij}} , \qquad (6.10)$$

---

[6] As for the transition matrix for a zeroth-order chain (i.e., independent trials).
[7] See Billingsley (1961), especially for the proof of Whittle's formula, Eq. (6.9).

where the left-hand side is simply the probability of realizing a particular sequence $\{x_1, x_2, \ldots, x_{n+1}\}$. For $i, j = 1, \ldots, s$, f_{ij} is the number of m, with $1 \leq m \leq n$, for which $a_m = i$ and $a_{m+1} = j$; F is therefore an $s \times s$ matrix, such that $\sum_{ij} f_{ij} = n$ and such that $f_{i\cdot} - f_{\cdot i} = \delta_{iu} - \delta_{iv}, i = 1, \ldots, s$, for some pair u, v, where $f_{i\cdot} = \sum_j f_{ij}$, and $\{f_{i\cdot}\}$ and $\{f_{\cdot j}\}$ are the frequency counts of $\{a_1, \ldots, a_n\}$ and $\{a_2, \ldots, a_{n+1}\}$, respectively, from which $f_{i\cdot} - f_{\cdot i} = \delta_{ia_1} - \delta_{ia_{n+1}}$. In Eq. (6.9), F^*_{uv} is the (v, u)th cofactor of the matrix $F^* = f^*_{ij}$, with components

$$f^*_{ij} = \begin{cases} \delta_{ij} - f_{ij}/f_{i\cdot} & \text{if } f_{i\cdot} > 0 \\ \delta_{ij} & \text{if } f_{i\cdot} = 0 . \end{cases} \tag{6.11}$$

Problem.   Prove that if P is stochastic, then any power of P is also stochastic.

The entropy of the transitions (i.e., the weighted variety of the transitions) can be found from each row of the stochastic matrix according to Eq. (2.5). The (informational) entropy of the process as a whole is then the weighted average of these entropies, the weighting being given by the equilibrium distribution of the states. Hence, in a sense the entropy of a Markov process is an average of averages.

Problem.   Consider the three-state Markov chain

$\rightarrow$	1	2	3
1	0.1	0.9	0.0
2	0.5	0.0	0.5
3	0.3	0.3	0.4

and calculate (i) the equilibrium proportions of the states 1, 2, and 3 and (ii) the average entropy of the entire process.

6.3   Random Walks

Consider an agent on a line susceptible to step right along the line with probability p and left with probability $q = 1 - p$. We can encode the walk by writing $+1$ for a right step and -1 for a left step. Many processes can be mapped onto the random walk (e.g., a nucleic acid sequence, with purines $\equiv -1$ and pyrimidines $\equiv +1$). If the walk is drawn in Cartesian coordinates as a polygon with the number of steps ("time") along the horizontal axis and the displacement along the vertical axis, then if s_k is the partial sum of the first k steps,

$$s_k - s_{k-1} = \pm 1, \qquad s_0 = 0, \qquad s_n = n(p - q) , \tag{6.12}$$

where n is the length of the path.

Definition.   Let $n > 0$ and x be integers. A path $(s_1, s_2, \ldots, s_n)$ from the origin to the point (n, x) is a polygonal line whose vertices have abscissae $0, 1, \ldots, n$ and ordinates $s_0, s_1, \ldots, s_n$ satisfying $s_k - s_{k-1} = \epsilon_k = \pm 1$, $s_0 = 0$, and $s_n = p - q$ (where p and q are now the numbers of symbols, $p + q = n$), with $s_n = x$.

There are 2^n paths of length n, but a path from the origin to an arbitrary point (n, x) exists only if n and x satisfy

$$n = n(p + q), \qquad x = n(p - q) .$$ (6.13)

In this case, the np positive steps can be chosen from among the n available places in

$$N_{n,x} = \binom{p + q}{p} = \binom{p + q}{q}$$ (6.14)

ways. The average distance travelled after n steps is $\sim n^{1/2}$, and the variance increases linearly with the number of steps.

Diffusion is an example of a random walk. The diffusivity (diffusion coefficient) D that gives the constant of proportionality in Fick's first and second laws[8] is given by λ^2/τ, where λ is the step length and τ is the duration of each step. The random walk is, of course, an example of a Markov chain.

Problem. Write out the Markovian transition matrix for a random walk in one dimension.

6.4   Noise

It might be thought that "noise" is the ultimate random, uncorrelated process. In reality, however, noise can come in various "colours" according to the exponent of its power spectrum.

---

[8]Fick's first law is

$$J_i = -D_i \nabla c_i ,$$ (6.15)

where J is the flux of substance i across a plane and c is its (position-dependent) concentration. In one dimension, this law simply reduces to $J = -D \partial c(x)/\partial x$, where x is the spatial coordinate. In most cases, especially in the crowded milieu of a living cell, it is more appropriate to use the (electro)chemical potential μ than the concentration, whereupon the law becomes

$$J_i = -D_i \nabla \mu_i (c_i/k_B T)$$ (6.16)

where T is the absolute temperature. Fick's second law, appropriate for time-varying concentrations, is

$$\partial c/\partial t = D \nabla^2 c .$$ (6.17)

If D itself changes with position (e.g., the diffusivity of a protein depends on the local concentration of small ions surrounding it), then we have

$$\partial c/\partial t = \nabla \cdot (D \nabla c) .$$ (6.18)

Let $x(t)$ describe a fluctuating quantity. It can be characterized by the two-point autocorrelation function

$$C_x(n) = \sum_{j=1}^{N} x_j x_{j-n} \qquad (6.19)$$

(in discrete form), where n is the position along a nucleic acid or protein sequence of N elements, and by the spectrum or amplitude spectral density

$$A_x(m) = \sum_{j=-\infty}^{\infty} x_j e^{-2\pi i m} , \qquad (6.20)$$

whose square is the power spectrum or power spectral density:

$$S_x(m) = |A_x(m)|^2 , \qquad (6.21)$$

where m is sequential frequency. The autocorrelation function and the power spectrum are just each other's Fourier transforms (the Wiener–Kintchin relations, applicable to stationary random processes).

A truly random process ["white noise," $w(t)$] should have no correlations in time. Hence,

$$C_w(\tau) \propto \delta(\tau) \qquad (6.22)$$

and

$$S_w(f) \propto 1 ; \qquad (6.23)$$

the power spectrum is convergent at low frequencies, but if one integrates up from some finite frequency toward infinity, one finds a divergence: there is an infinite amount of power at the highest frequencies; that is, a plot of $w(t)$ is infinitely choppy and the instantaneous value of $w(t)$ is undefined!

White noise is also called Johnson (who first measured it experimentally, in 1928) or Nyquist (who first derived its power spectrum theoretically) noise. It is characteristic of the voltage across a resistor measured at open circuit and is due to the random motions of the electrons. The integral of white noise,

$$B(t) = \int w(t)\, dt , \qquad (6.24)$$

corresponds to a random walk or Brownian motion (hence, "brown noise"). Its power spectrum is

$$S_B(f) \propto 1/f^2 ; \qquad (6.25)$$

that is, it is convergent when integrating to infinity, but divergent when integrating down to zero frequency. In other words, the function has a well-defined value at each point, but wanders ever further from its initial value at longer and longer times; that is, it does not have a well-defined mean value.

If current is flowing across a resistor, then the power spectrum of the voltage fluctuations $S_F(f) \propto 1/f$ ["$1/f$ noise," sometimes called "fractional Gaussian noise" (FGN)], as a special case of fractionally integrated white noise. FGNs are characterized by a parameter F: the mean distance travelled in the process described by its integral $G_F(t) = \int x_F(t)\, dt$ is proportional to t^F, and the power spectrum

$S_G(f) \propto 1/f^{2F-1}$. White noise has $F = \frac{1}{2}$, and $1/f$ noise has $F = 1$. It is divergent when integrated to infinite frequency and when integrated to zero frequency, but the divergences are only logarithmic. $1/f$ noise exhibits very long-range correlations, the physical reason for which is still a mystery. Many natural processes exhibit $1/f$ noise.

6.5   Complexity

The notion of complexity occurs rather frequently in biology, where one often refers to the complexity of this or that organism (cf. *biological complexity*, Sect. 9.7). Several procedures for ascribing a numerical value to it have been devised, but for all that it remains somewhat elusive. When we assert that a mouse is more complex than a bacterium (or than a fly), what do we actually mean? Intuitively, the assertion is unexceptionable—most people would presumably readily agree that man is the most complex organism of all. Is our genome the biggest (as may once have been believed)? No. Do we have more cell types than other organisms? Yes, and the mouse has more than the fly, but then complexity becomes merely a synonym for variety. Or does it reflect what we can do? Man alone can create poems, theories, musical compositions, paintings, and so forth. However, although one could perhaps compare the complexity of different human beings on that basis, it would be useless for the rest of the living world. Is complexity good or bad? A complex theory that nobody apart from its inventor can understand might be impressive, but not very useful. On the other hand, we have the notion, again rather intuitive, that a complex organism is more adaptable than a simple one, because it has more possibilities for action; hence, it can better survive in a changing environment.[9]

Other pertinent questions are whether complexity is an absolute attribute of an object, or does it depend on the level of detail with which one describes it (in other words, how its description is encoded—an important consideration if one is going to extract a number to quantify complexity)? Every writer on the subject seems to introduce his own particular measure of complexity, with a corresponding special name—what do these different measures have in common? Do printed copies of a Shakespeare play have the same complexity as the original manuscript? Does the fiftieth edition have less complexity than the first?

The antonym of complexity is simplicity; the antonym of randomness is regularity. A highly regular pattern is also simple. Does this, then, suggest that complexity is a synonym for randomness?

An important advance was Kolmogorov's notion of algorithmic complexity (also called algorithmic information content or AIC) as a criterion for randomness. As we

---

[9]If this is so, it then seems rather strange that so much ingenuity is expended by presumably complex people to make their environments more uniform and unchanging, in which case they will tend to lose their competitive advantage.

have seen near the beginning of this chapter (footnote 4), the AIC, $K(s)$, of a string s is the length of the smallest program (running on a universal computing machine) able to print out s. Henceforth we shall mainly consider the complexity of strings (objects can, of course, be encoded as strings). If there are no regularities, $K(s)$ will have its maximum possible value, which will be roughly equal to the length of the string; no compression is possible and the string has to be printed out verbatim.[10] Hence,

$$K_{\max} = |s| . \tag{6.26}$$

Any regularities (i.e., constraints in the choice of successive symbols) will diminish the value of K. We call $K_{\max}$ the unconditional complexity; it is actually a measure of regularity.

This definition leads to the intuitively unsatisfying consequence that the highest possible complexity, the least regularity, the greatest potential information gain, etc. is possessed by a purely random process, which then implies that the output of the proverbial team of monkeys tapping on keyboards is more complex than a Shakespeare play (the difference would, however, vanish if the letters of the two texts were encoded in such a way that the same symbol was used to encode each letter). What we would like is some quantity that is small for highly regular structures (low disorder), then increases to a maximum as the system becomes more disordered, and finally falls back to a low value as the disorder approaches pure randomness.

In order to overcome this difficulty, Gell-Mann has proposed *effective complexity* to be proportional to the length of a concise description of a set of an object's regularities, which amounts to the algorithmic complexity of the description of the set of regularities. This prescription certainly fulfils the criterion of correspondence with the intuitive notion of complexity; both a string consisting of one type of symbol and the monkey-text would have no variety in their regularity and hence minimal complexity. One way of assessing the regularities is to divide the object into parts and examine the mutual algorithmic complexity between the parts. The effective complexity is then proportional to the length of the description of those regularities.

Correlations within a symbolic sequence (string) have been used by Grassberger (1986) to define effective measure complexity (EMC) from the correlation information (see Sect. 2.2):

$$\eta = \sum_{m=2}^{\infty} (m - 1)k_m . \tag{6.27}$$

In effect, it is a weighted, average correlation length.

A more physically oriented approach has been proposed by Lloyd and Pagels (1988). Their notion of (thermodynamic) depth attempts to measure the process

---

[10]Many considerations of complexity may be reduced to the problem of printing out a number. Thus, the complexity of a protein structure is related to the number specifying the positions of the atoms, or dihedral angles of the peptide groups, which is equivalent to selecting one from a list of all possible conformations; the difficulty of doing that is roughly the same as that of printing out the largest number in that list.

whereby an object is constructed. A complex object is one that is difficult to put together;[11] the average complexity of a state is the Shannon entropy of the set of trajectories leading to that state $(-\sum p_i \log p_i$, where p_i is the probability that the system has arrived at that state by the ith trajectory) and the depth $\mathcal{D}$ of a system in a macroscopic state d is $\sim -\log p_i$. An advantage of this process-oriented formulation is the way in which the complexity of copies of an object can be dealt with; the depth of a copy, or any number of copies, is proportional to the depth of making the original object plus the depth of the copying process.

Process is used by Lempel and Ziv (1976) to derive a complexity measure, called production complexity, based on the gradual buildup of new patterns (rate of vocabulary growth) along a sequence s:

$$\mathfrak{c}(s) = \min\{c_H(s)\} \tag{6.28}$$

where minimization is over all possible histories of s and $c_H(s)$ is the number of components in the history. The production history $H(s)$ is defined as the parsing of s into its m components (words):

$$H(s) = s(1, h_1)s(h_1 + 1, h_2) \cdots s(h_{m-1} + 1, h_m) . \tag{6.29}$$

$\mathfrak{c}(s)$ is thus the least possible number of steps in which s can be generated according to the given rules of production.

In order to go beyond purely internal qualities (i.e., correlations) of the string, it will be useful to introduce some additional quantities, such as the joint algorithmic complexity $K(s, t)$, the length of the smallest program required to print out two strings s and t:

$$K(s, t) \approx K(t, s) \lesssim K(s) + K(t) ; \tag{6.30}$$

the mutual algorithmic information

$$K(s : t) = K(s) + K(t) - K(s, t) \tag{6.31}$$

(which reflects the ability of a string to share information with another string); conditional algorithmic information (or conditional complexity)

$$K(s|t) = K(s, t) - K(t) \tag{6.32}$$

(i.e., the length of the smallest program that can compute s from t); and algorithmic information distance

$$D(s, t) = K(s, t) + K(t|s) \tag{6.33}$$

(the reader may verify that this measure fulfils the usual requirements for a distance).

Adami and Cerf (2000) have emphasized that randomness and complexity only exist with respect to a specific, defined, environment e (i.e., context). Consider the conditional complexity $K(s|e)$. The smallest program for computing s from e will only contain elements unrelated to e, since if they were related, they could be obtained

---

[11]Cf. the nursery rhyme *Humpty Dumpty sat on a wall/Humpty Dumpty had a great fall/And all the king's horses and all the king's men/Couldn't put Humpty together again.* It follows that Humpty Dumpty had great depth, hence complexity.

(i.e., deduced) from e with a program tending to size zero. Hence, $K(s|e)$ quantifies those elements in s that are random (with respect to e).[12] In principle, we can now use the mutual algorithmic information defined by Eq. (6.31) to determine

$$K(s : e) = K_{max} - K(s|e) , \qquad (6.34)$$

which represents the number of meaningful elements in string s, although it might not be practically possible to compute $K(s|e)$ unless one is aware of the coding scheme whereby some of e is encapsulated in s. A possible way of overcoming this difficulty is opened where there exist multiple copies of a sequence that have adapted independently to e. It may then reasonably be assumed that the coding elements are conserved (and have a nonuniform probability distribution), whereas the noncoding bits are fugitive (and have a uniform probability distribution). The information about e contained in the ensemble S of copies is then the Shannon index $I(S) - I(S|e)$. In finite ensembles, the quantity

$$I(S|e) = - \sum_s p(s|e) \log p(s|e) \qquad (6.35)$$

can be estimated by sampling the distribution $p(s|e)$.

Computational complexity reflects how the number of elementary operations required to compute a number increases with the size of that number. Hence, the computational complexity of "011011011011011011 ..." is of order unity, since one merely has to specify the number of repetitions.

Algorithmic and computational complexity are combined in the concept of logical depth,[13] defined as the number of elementary operations (machine cycles) required to calculate a string from the shortest possible program. Hence, the number π, whose specification requires only a short program, has considerable logical depth because that program has to execute many operations to yield π.

Problem. A deep notion is generally held to be more meaningful than a shallow one. Could one, then, identify complexity with meaning? Discuss the use of the ways of quantifying complexity, especially effective complexity, as a measure of meaning (cf. Sect. 2.3.2).

References

Adami C, Cerf NJ (2000) Physical complexity of symbolic sequences. Phys D 137:62–69

Bennett CH (1988) Logical depth and physical complexity. In: Herken R (ed) The Universal Turing Machine—A Half-Century Survey. Oxford, University Press, pp 227–257

Billingsley P (1961) Statistical methods in Markov chains. Ann Math Statist 32:12–40

---

[12]If there is no environment, then all strings have the maximum complexity, K_{max}.
[13]Due to Bennett (1988).

Grassberger P (1986) Toward a quantitative theory of self-generated complexity. Int J Theor Phys
    25:907–938
Lempel A, Ziv J (1976) On the complexity of finite sequences. IEEE Trans Info Theory IT–22:75–81
Lloyd S, Pagels H (1988) Complexity as thermodynamic depth. Ann Phys 188:186–213
van der Waerden BL (1927) Beweis einer Baudet'schen Vermutung. Nieuw Arch Wiskunde 15:212–
    216
Volchan SB (2002) What is a random sequence? Am Math Monthly 109:46–63

Systems, Networks, and Circuits

Just as we are often interested in events that are composed of many elementary (simple) events, in biology the objects under scrutiny are vastly complex objects composed of many individual molecules (the molecule is probably the most appropriate level of coarse graining for the systems we are dealing with). Since these components are connected together, they constitute a system. The essence of a system is that it cannot be usefully decomposed into its constituent parts. More formally, following R.L. Ackoff we can assert that two or more objects (which may be entities, or activities, etc.) constitute a *system* if the following four conditions are satisfied:

1. One can talk meaningfully of the behaviour of the whole of which they are the only parts;
2. The behaviour of each part can affect the behaviour of the whole;
3. The way each part behaves and the way its behaviour affects the whole depends on the behaviour of at least one other part;
4. No matter how one subgroups the parts, the behaviour of each subgroup will affect the whole and depends on the behaviour of at least one other subgroup.

There are various corollaries, one of the most important and practical of which is that a system cannot be investigated by looking at its components individually, or by varying one parameter at a time, as Fisher (1951) seems to have been the first to realize. Thus, a *modus operandi* of the experimental scientist inculcated at an early age and reinforced by the laboratory investigation of "simple systems"[1] turns out to be inappropriate and misleading when applied to most phenomena involving the living world.

---

[1] Here we plead against the use of the terms "simple system" and "complex system": the criteria given above imply that no system is simple, and that every system is complex.

© Springer-Verlag London 2015
J. Ramsden, *Bioinformatics*, Computational Biology 21,
DOI 10.1007/978-1-4471-6702-0_7

Another corollary is that the concept of feedback, which is usually clear enough to apply to two-component systems, is practically useless in more complex systems.[2]

In this chapter, we shall first consider the approach of general systems theory, largely pioneered by Bertalanffy (1993). This allows some insight into the behaviour of very simple systems with not more than two components, but thereafter statistical approaches have to be used.[3] This is successful for very large systems, in which statistical regularities can be perceived; the most difficult cases are those of intermediate size. Some properties of networks *per se* will then be examined, followed by a brief look at synergetics (systems with a diffusion term), and the final section deals with complex evolving systems.

Problem. Consider various familiar objects, and ascertain using the above criteria whether they are systems.

7.1   General Systems Theory

Consider a system containing n interacting elements $G_1, G_2, \ldots, G_n$. Let the values of these elements be $g_1, g_2, \ldots, g_n$. For example, if the G denote species of animals, then g_1 could be the number of individual animals of species G_1. The temporal evolution of the system is then described by

$$\frac{dg_1}{dt} = \mathcal{G}_1(g_1, g_2, \ldots, g_n)$$

$$\frac{dg_2}{dt} = \mathcal{G}_2(g_1, g_2, \ldots, g_n)$$

$$\vdots \tag{7.1}$$

$$\frac{dg_n}{dt} = \mathcal{G}_n(g_1, g_2, \ldots, g_n)$$

where the functions $\mathcal{G}$ include terms proportional to $g_1, g_1^2, g_1^3, \ldots, g_1g_2, g_1g_2g_3$, etc. In practice, many of the coefficients of these terms will be close or equal to zero.

If we only consider one variable,

$$\frac{dg_1}{dt} = \mathcal{G}_1(g_1) . \tag{7.2}$$

Expanding gives

$$\frac{dg_1}{dt} = rg_1 - \frac{r}{K}g_1^2 + \cdots \tag{7.3}$$

---

[2]Even in two component systems its nature can be elusive. For example, as Ashby (1956) has pointed out, are we to speak of feedback between the position and momentum of a pendulum? Their interrelation certainly fulfils all the formal criteria for the existence of feedback.

[3]Robinson (1998) has recently proved that all possible chaotic dynamics can be approximated in only three dimensions.

where $r > 0$ and $K > 0$ are constants. Retaining terms up to g_1 gives simple exponential growth,

$$g_1(t) = g_1(0)e^{rt} \tag{7.4}$$

where $g_1(0)$ is the quantity of g_1 at $t = 0$. Retaining terms up to g_1^2 gives

$$g_1(t) = \frac{K}{1 + e^{-r(t-m)}}, \tag{7.5}$$

the so-called logistic equation, which is sigmoidal with a unique point of inflexion at $t = m$, $g_1 = K/2$ at which the tangent to the curve is r, and asymptotes $g_1 = 0$ and $g_1 = K$. r is called the growth rate and K is called the carrying capacity in ecology.[4]

Consider now two objects,

$$\left. \begin{aligned} dg_1/dt &= a_{11}g_1 + a_{12}g_2 + a_{111}g_1^2 + \cdots \\ dg_2/dt &= a_{21}g_1 + a_{22}g_2 + a_{211}g_1^2 + \cdots \end{aligned} \right\} \tag{7.6}$$

in which the functions $\mathcal{G}$ are now given explicitly in terms of their coefficients a (a_{11}, for example, gives the time in which an isolated G_1 returns to equilibrium after a perturbation). The solution is

$$\left. \begin{aligned} g_1(t) &= g_1^* - h_{11}e^{\lambda_1 t} - h_{12}e^{\lambda_2 t} - h_{111}e^{2\lambda_1 t} - \cdots \\ g_2(t) &= g_2^* - h_{21}e^{\lambda_1 t} - h_{22}e^{\lambda_2 t} - h_{211}e^{2\lambda_1 t} - \cdots \end{aligned} \right\} \tag{7.7}$$

where the starred quantities are the stationary values, obtained by setting $dg_1/dt = dg_2/dt = 0$, and the λs are the roots of the characteristic equation, which is (ignoring all but the first two terms of the right hand side of Eq. 7.6)

$$\begin{vmatrix} a_{11} - \lambda & a_{12} \\ a_{21} & a_{11} - \lambda \end{vmatrix} = 0. \tag{7.8}$$

Depending on the values of the a coefficients, the phase diagram (i.e. a plot of g_1 vs g_2) will tend to a point (all λ are negative), or a limit cycle (the λ are imaginary, hence there are periodic terms), or there is no stationary state (λ are positive). Regarding the last case, it should be noted that however large the system, a single positive λ will make one of the terms in (7.7) grow exponentially and hence rapidly dominate all the other terms.

Although this approach can readily be generalized to any number of variables, the equations can no longer be solved analytically and indeed the difficulties become forbidding. Hence one must turn to statistical properties of the system. Equation (7.6) can be written compactly as

$$\dot{\mathbf{g}} = A\mathbf{g} \tag{7.9}$$

---

[4]Unrelated to the previous K (Sects. 2.1.3 and 6.5). We retain the same symbol here because of "K-selection" (Sect. 10.8.4), an expression well anchored in the literature of ecology. Yet another unrelated use of K is in Kauffman's (1984) NK model (Sect. 7.2.3).

where **g** is the vector $(g_1, g_2, \ldots)$, $\dot{\mathbf{g}}$ its time differential, and A the matrix of the coefficients a_{11}, a_{12} etc. connecting the elements of the vector. The binary connectivity C_2 of A is defined as the proportion of nonzero coefficients.[5] In order to decide whether the system is stable or unstable, we merely need to ascertain that none of the roots of the characteristic equation are positive, for which the Routh–Hurwitz criterion can be used without actually having to solve the equation. Gardner and Ashby (1970) determined the dependence of the probability of stability on C_2 by distributing nonzero coefficients at random in the matrix A for various values of the number of variables n. They found a sharp transition between stability and instability: for $C < 0.13$, a system will almost certainly be stable, and for $C > 0.13$, almost certainly unstable. For very small n the transition became rather gradual, viz. for $n = 7$ the probability of stability is 0.5 at $C_2 \approx 0.3$, and for $n = 4$, at $C_2 \approx 0.7$.

7.1.1 Automata

We can generalize the Markov chains from Sect. 6.2 by writing Eq. (7.9) in discrete form:

$$\mathbf{g}' = A\mathbf{g} \qquad (7.10)$$

i.e. the transformation A is applied at discrete intervals and $\mathbf{g}'$ denotes the values of g at the epoch following the starting one. The value of g_i now depends not only on its previous value, but also on the previous values of some or all of the other $n - 1$ components. Generalizations to the higher-order coefficients are obvious but difficult to write down; we should bear in mind that application of this approach to the living cell is likely to require perhaps third or fourth order coefficients, but that the corresponding matrices will be extremely sparse.

The analysis of such systems usually proceeds by restricting the values of the g to integers, and preferably to just zero or one (Boolean automata). Consider an automaton with just three components, each of which has an output connected to the other two. Equation (7.10) becomes

$$\begin{pmatrix} g_1 \\ g_2 \\ g_3 \end{pmatrix}' = \begin{pmatrix} 0\ 1\ 1 \\ 1\ 0\ 1 \\ 0\ 0\ 1 \end{pmatrix} \diamond \begin{pmatrix} g_1 \\ g_2 \\ g_3 \end{pmatrix} \qquad (7.11)$$

where $\diamond$ denotes that the additions in the matrix multiplication are to be carried out using Boolean AND logic; i.e., according to Table 7.1. Enumerating all possible starting values leads to the state structure shown in Fig. 7.1. The problem at the end of this subsection will help the reader to be convinced that state structure is not closely related to physical structure (the pattern of interconnexions). In fact, to study a system one needs to determine the state structure and know both the interconnexions and the functions of the individual objects (cells).

---

[5]The ternary connectivity takes into account connexions between three elements, i.e. contains coefficients like a_{123}, etc.

Table 7.1 Truth table for an AND gate	Input	Output
	0	0
	1	0
	2	1

Fig. 7.1 State structure of the automaton represented by Eq. (7.11) and Table 7.1

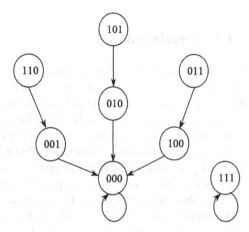

Most of the work on the evolution of automata (state structures) considers the actual structure (interconnexions) and the individual cell functions to be immutable. For biological systems, this appears to be an oversimplification. Relative to the considerable literature on the properties of various kinds of networks, very little has been done on evolving networks, however.[6]

Problem. Determine the state structure of an automaton if (i) the functions of the individual cells are changed from those represented by (7.11) such that G_1 becomes 1 whenever G_2 is 1, G_2 becomes 1 whenever G_3 is 1, and G_3 becomes 1 whenever G_1 and G_2 have the same value; (ii) keep these functions, but connect G_1's output to itself and G_3, G_2's output to itself, G_1 and G_3, and G_3's output to G_2; and (iii) keep these new interconnexions, but restore the functions to those represented by (7.11) and Table 7.1. Compare the results with each other and with Fig. 7.1.

7.1.2   Cellular Automata

This term is usually applied to cells arranged in spatial proximity to each other, whose states are updated according to a rule such as $n_i' = (n_{i-1} + n_i + n_{i+1})$ mod 2, where n_i is the current state of the ith cell. The most widely studied ones are only

---

[6]An exception is Érdi and Barna's (1984) work on a model of neuron interconnexions, simulating Hebb's rule (traffic on a synapse strengthens it, i.e. increases its capacity).

connected to their nearest neighbours. Despite this simplicity, their evolution can be rather elaborate and even unpredictable. Wolfram (1983) has made an exhaustive study of one dimensional cellular automata, in which the cells are arranged on a line. Higher dimensional automata are useful in analysing biological processes; for example a two-dimensional automaton can be used to investigate neurogenesis in a membrane of undifferentiated precursor cells.[7]

7.1.3 Percolation

Consider a spatial array of at least two dimensions, with cells able to take values of zero or one, signifying respectively "impermeable" and "permeable" to some agent migrating across the array and only able to move from a permeable site to a nearest neighbour that is also permeable. Let p be the probability that a cell has the value 1. If ones are sparse (i.e. low p), the mobility of the agent will be restricted to small isolated islands. The most important problem in this field is to determine the mean value of p at which an agent can span the entire array via its nearest neighbour connexions. This is so-called "site percolation".[8]

A possible approach to determine the critical value p_c is as follows: the probability that a single permeable cell on a square lattice is surrounded by impermeable ones (i.e. is a singlet) is pq^4, where $q = 1 - p$. Defining $n_s(p)$ to be the average number of s-clusters per cell, then we have $n_2(p) = 2p^2q^6$ for doublets, (the factor 2 arises because of the two possible perpendicular orientations of the doublet), $n_3(p) = 2p^3q^8 + 4p^3q^7$ for triplets (linear and bent), etc. If there are few permeable cells, $\sum_s sn_s(p) = p$; if there are many we can expect most of the particles to belong to an infinite (in the limit of an infinite array) cluster, hence $\sum_s sn_s(p) + P_\infty = p$, and the mean cluster size $S(p) = \sum_s s^2n_s(p)/p$. If $S(p)$ is now expanded in powers of p, one finds that at a certain value of p the series diverges; this is when the infinite (spanning) cluster appears, and we can call the array "fully connected". The remarkable Galam and Mauger (1996) formula gives this critical threshold p_c for isotropic lattices:

$$p_c = a[(D - 1)(C - 1)]^{-b} \qquad (7.12)$$

where D is the dimension, C the connectivity of the array (i.e. the number of nearest neighbours of any cell), and a and b are constants with values 1.2868 and 0.6160 respectively, allows one to calculate the critical threshold for many different types of network.

---

[7]Luthi et al. (1998).

[8]In "bond percolation" movement occurs along links joining nearest neighbours with probability p. Every bond process can be converted into a site one, but not every site process is a bond one.

Fig. 7.2 A fragment of a
network (graph). Note the two
types of nodes and that some
of the vertices are directed

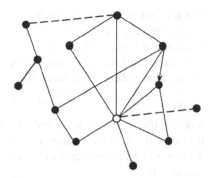

7.2 Networks (Graphs)

The cellular automata considered above (Sect. 7.1.2) are examples of *regular* networks (of automata): the objects are arranged on a regular lattice and connected in an identical fashion with neighbours. Consider now a collection of objects (nodes or vertices) characterized by number, type and interconnexions (edges or links). Figure 7.2 represents an archetypical fragment of a network (graph). The connexions between nodes can be given by an adjacency matrix A whose elements a_{ij} give the strength of the connexion (in a Boolean network $a = 1$ or 0, respectively connexion present or absent) between nodes i and j. In a directed graph A need not be symmetric. An oriented graph is a directed graph in which every edge is oriented. The element $[A^r]_{ij}$ gives the number of walks of length r between nodes i and j.[9]

If the only knowledge one has is of the positions of the objects in space, the adjacency matrix can be constructed by defining an edge to exist between a pair of objects if the distance between them is less than a certain threshold.

We begin by considering the structural properties of a network. Useful parameters are the following: N, the number of nodes; E, the number of edges; $\langle k \rangle$, the average degree of each node (the number of other vertices to which a given vertex is joined); L, the network diameter (the smallest number of edges connecting a randomly chosen pair of nodes; this is a global property);[10] and the cliquishness $\mathfrak{C}$ defined as the fraction of nodes linked to a given vertex that are themselves connected (this is a local property), or, in other words, the (average) number of times any two nodes connected to a third node are themselves connected. Hence, this is equivalent to the number of closed triangles in the network, that is,

$$\mathfrak{C} \propto \mathrm{Tr}\, A^3 , \tag{7.13}$$

from which a relative clustering coefficient can be defined as

$$\mathfrak{C}_r = \mathfrak{C}/N . \tag{7.14}$$

---

[9] A mesh network is one in which there are at least two pathways of communication to each node. Such networks are, of course, more resilient with respect to failure of some pathways.

[10] A useful way to compute L is given by Raine and Norris (2002).

The maximum number of possible edges in a network is $N(N-1)/2$ (the factor 2 in the denominator arising because each edge has two endpoints); the connectivity C is the actual number of edges (which may be weighted by the strength of each edge) divided by the maximum number. A graph with $C = 1$ is known as complete. The degree matrix D is constructed as

$$D = \text{diag}(k_1, \ldots, k_N) \,, \tag{7.15}$$

where k_i the degree of the ith node, from which the the Laplace matrix $L = D - A$ and the normalized Laplace matrix $\bar{L} = I - D^{-1}A$ can be determined. The eigenvalues of L are useful for giving rapid information about the connectivity, robustness, stability, and so forth.

Two important generic topologies of graphs are as follows:

(i) *random (Erdős–Rényi) graphs*. Each pair of nodes is connected with probability p; the connectivity of such a network peaks strongly at its average value and decays exponentially for large connectivities. The probability $p(k)$ that a node has k edges is given by $\mu^k e^{-\mu}/k!$, where $\mu = 2Np$ is the mean number of edges per node. The smallest number of edges connecting a randomly chosen pair of nodes (i.e., the network diameter L) is $\sim \log N$ (cf. $\sim N$ for a regular network). The cliquishness (clustering coefficient) $\mathfrak{C} = \mu$. This type of graph has a percolation-like transition. If there are M interconnexions, then when $M = N/2$ a giant cluster of connected nodes appears.

A special case of the random graph is the small world. This term applies to networks in which the smallest number of edges connecting a randomly chosen pair of nodes is comparable to the $\log N$ expected for a random network (i.e., much smaller than for a regular network), whereas the local properties are characteristic of a regular network (i.e., the clustering coefficient is high). The name comes from the typical response, "It's a small world!" uttered when it turns out that two people meeting for the first time and with no obvious connexion between them have a common friend.[11]

(ii) *the "scale-free" networks*, in which the probability $P(k)$ of a node having k links $\sim k^{-\gamma}$, where γ is some constant.[12] A characteristic feature of a scale-free network is therefore that it possesses a very small number of highly connected nodes. Many properties of the network are highly vulnerable to the removal of these nodes.

---

[11] The first published account appears in F. Karinthy, Láncszemek (in: *Címszavak a Nagy Enciklopédiához*, vol. 1, pp. 349–354. Budapest: Szépirodalmi Könyvkiadó (1980). It was first published in the 1920s). A simple way of constructing a model small-world network has been given by Watts and Strogatz (1998): start with a ring of nodes each connected to their k nearest neighbours (i.e., a regular network). Then detach connexions from one of their ends with probability p and reconnect the freed end to any other node (if $p = 1$, then we recover a random network). As p increases, L falls quite rapidly, but $\mathfrak{C}$ only slowly (as $3(\mu - 2)/[4(\mu - 1)]$). The small-world property applies to the régime with low L but high $\mathfrak{C}$.

[12] Scale-free networks seem to be widespread in the world. The first systematic investigation of their properties is supposed to have been conducted by Dominican monks in the thirteenth and fourteenth centuries, in connexion with eradicating heresy.

A simple algorithm for generating scale-free networks was developed by Albert and Barabási (2002): Start with a small number m_0 of nodes and add, stepwise, new nodes with $m (\leq m_0)$ edges, linking each new node to m existing nodes. Unlike the random addition of edges that would result in an Erdős–Rényi graph, the nodes are preferentially attached to already well-connected nodes; that is, the probability that a new node will be connected to existing node i is

$$P(k_i) = k_i / \sum_j k_j \; . \tag{7.16}$$

After t steps, one has $m_0 + t$ nodes and mt edges, and the exponent γ appears (from numerical simulations) to be 3.

The average degree of this network remains constant as it grows. Empirical studies have shown, however, that in many natural systems, the average degree increases with growth (this phenomenon is called "accelerated growth"); in other words, each new node is connected to a fixed fraction of the existing nodes. In this case, $E \sim N^2$.

7.2.1 Trees

A tree is a graph in which each pair of vertices is joined by a unique edge; there is exactly one more vertex than the number of edges. In a binary tree, each vertex has either one or three edges connected to it. A rooted tree has one particular node called the root (corresponding to the point at which the trunk of a real (biological) tree emerges from the ground). Trees represent ultrametric space satisfying the strong triangle inequality

$$d(x, z) \leq \max\{d(x, y), d(y, z)\} \; , \tag{7.17}$$

where x, y, and z are any three nodes and d is the distance between a pair of nodes. Trees are especially useful for representing hierarchical systems. The clustering coefficient of a tree equals zero.

The complexity $\mathcal{C}$ of a tree T consisting of b subtrees $T_1, \ldots, T_b$ (i.e., b is the number of branches at the root), of which k are not isomorphic, is defined as[13]

$$\mathcal{C} = \mathcal{D} - 1 \; , \tag{7.18}$$

where the diversity measure $\mathcal{D}$ counts both interactions between subtrees and within them and is given by

$$\mathcal{D} = (2^k - 1) \prod_{j=1}^{k} \mathcal{D}(T_j^{(i)}) \; . \tag{7.19}$$

If a tree has no subtrees, $\mathcal{D} = 1$; the complexity of this, the simplest kind of tree, is set to zero (hence, Eq. 7.18). Any tree with a constant branching ratio at each mode will also have $\mathcal{D} = 1$ and, hence, zero complexity. This complexity measure satisfies the intuitive notion that the most complex structures are intermediate between regular and random ones (cf. Sect. 6.5).

---

[13] See Huberman and Hogg (1986).

7.2.2   Complexity Parameters

There are various measures of network complexity:

1. κ, the number of different spanning trees of the network
2. Structural complexity, the number of parameters needed to define the graph
3. Edge complexity, the variability of the second shortest path between two nodes
4. Network or β-complexity, given by the ratio $\mathfrak{C}/L$
5. Algorithmic complexity, the length of the shortest algorithm needed to describe the network (see also Chap. 6).

7.2.3   Dynamical Properties

The essential concepts of physical structure and state structure were already introduced in Sect. 7.1.1 and Fig. 7.1. A considerable body of work has been accomplished along these lines: investigating the state structures of simple, or simply constructed, networks. Kauffman (1984), in particular, has studied large randomly connected Boolean networks, with the interesting result that if each node has on average two inputs from other nodes; typically, the state structure comprises about $N^{1/2}$ cyclic attractors, where N is the number of nodes (i.e., far fewer than the 2^N potentially accessible states).

More generally, Kauffman (1984) considered strings of N genes, each present in the form of either of two alleles (0 and 1).[14] In the simplest case, each gene is independent, and when a gene is changed from one allele to the other, the total fitness changes by at most $1/N$. If epistatic interactions (when the action of one gene is modified by others) are allowed, the fitness contribution depends on the gene plus the contributions from K other genes (the NK model),[15] and the fitness function or "landscape" becomes less correlated and more rugged.[16]

Érdi and Barna (1984) have studied how the pattern of connexions changes when their evolution is subjected to certain simple rules; the evolution of networks of automata in which the properties of the automata themselves can change has barely been touched, although this, the most complex and difficult case, is clearly the one closest to natural networks within cells and organisms. The study of networks and their application to real-world problems has, in effect, only just begun.

---

[14] Here we preempt some of the discussion in Sect. 10.9.2.

[15] This is yet another use of the symbol K—see footnote 4 earlier in this chapter.

[16] Note that, as pointed out by Jongeling (1996), fitness landscapes cannot be used to model selection processes if the entities being selected do not compete.

7.3 Synergetics

General systems theory (Sect. 7.1) can be further generalized and made more powerful by including a diffusion term:

$$\frac{\partial u_i}{\partial t} = \frac{1}{\tau_i} F_i(u_1, u_2, \ldots, u_n) + D_i \Delta u_i, \qquad i = 1, 2, \ldots, n . \tag{7.20}$$

u_i is a dynamic variable (e.g., the concentration of the ith object at a certain point in space), $F_i(u_i)$ are functions describing the interactions, τ_i is the characteristic time of change, and D_i is the diffusion coefficient (diffusivity) of the ith object. Equation (7.20) is thus a reaction–diffusion equation that explicitly describes the spatial distribution of the objects under consideration. The diffusion term tends to zero if the diffusion length $l_i > L$, the spatial extent of the system, where

$$l_i = D_i^{1/2} \tau_i . \tag{7.21}$$

Although solutions of Eq. (7.20) might be difficult for any given case under explicit consideration, in principle we can use it to describe any system of interest. This area of knowledge is called synergetics. Note that the "unexpected" phenomena often observed in elaborate systems can be easily understood within this framework, as we shall see.

One expects that the evolution of a system is completely described by its n equations of the type (7.20), together with the starting and boundary conditions. Suppose that a stationary state has been reached, at which all of the derivatives are zero, and described by the variables $\bar{u}_1, \ldots, \bar{u}_n$, at which all the functions F_i are zero. Small deviations δu_i may nevertheless occur and can be described by a system of linear differential equations

$$\frac{d}{dt} \delta u_i = \sum_j^n a_{ij} \delta u_j , \tag{7.22}$$

where the coefficients a_{ij} are defined by

$$a_{ij} = \frac{\partial F_i}{\partial u_i}\bigg|_{u_i = \bar{u}_i} . \tag{7.23}$$

The solutions of Eq. (7.22) are of the form

$$\delta u_j(t) = \sum_j^n \epsilon_{ij} e^{\lambda_i t} , \tag{7.24}$$

where the ϵ_{ij} are coefficients proportional to the starting deviations [viz. $\epsilon = \delta u(0)$]. The λs are called the Lyapunov numbers, which can, in general, be complex numbers, the eigenvalues of the system; they are the solutions of the algebraic equations

$$\det|a_{ij} - \delta_{ij}\lambda_j| = 0 , \tag{7.25}$$

where δ_{ij} is Kronecker's delta.[17] We emphasize that the Lyapunov numbers are purely characteristic of the system; that is, they are not dependent on the starting

---

[17] $\delta_{ij} = 0$ when $i \neq j$ and 1 when $i = j$.

conditions or other external parameters—provided the external influences remain small.

If all of the Lyapunov numbers are negative, the system is stable—the small deviations decrease in time. On the other hand, if at least one Lyapunov number is positive (or, in the case of a time-dependent Lyapunov number, if the real part becomes positive as time increases), the system is unstable, the deviations increase in time, and this is what gives rise to "unexpected" phenomena. If none are positive, but there are some zero or pure imaginary ones, then the stationary state is neutral.

7.3.1   Some Examples

The simplest bistable system is described by

$$\frac{du}{dt} = u - u^3 . \tag{7.26}$$

There are three stationary states, at $u = 0$ (unstable; the Lyapunov number is $+1$) and $u = \pm 1$ (both stable), for which the equation for small deviations is

$$\frac{d}{dt}\delta u = -3\delta u \tag{7.27}$$

and the Lyapunov numbers are -3. This system can be considered as a memory box with an information volume equal to $\log_2$(number of stable stationary states) = 1 bit.

A slightly more complex system is described by the two equations

$$\left.\begin{array}{l} du_1/dt = u_1 - u_1 u_2 - a u_1^2 \\ du_2/dt = u_2 - u_1 u_2 - a u_2^2 \end{array}\right\} . \tag{7.28}$$

The behaviour of such systems can be clearly and conveniently visualized using a phase portrait (e.g., Fig. 7.3). To construct it, one starts with arbitrary points in the (u_1, u_2) plane and uses the right-hand side of Eq. (7.28) to determine the increments. The main isoclines (at whose intersections the stationary states are found) are given by

$$\left.\begin{array}{l} du_1/dt = F_1(u_1, u_2) = 0 \\ du_2/dt = F_2(u_1, u_2) = 0 \end{array}\right\} . \tag{7.29}$$

Total instability, in which every Lyapunov number is positive, results in dynamic chaos. Intermediate systems have strange attractors (which can be thought of as stationary states smeared out over a region of phase space rather than contracted to a point), in which the chaotic régime occurs only in some portions of phase space.

7.3.2   Reception and Generation of Information

If the external conditions are such that in the preceding example (Eq. 7.28) the starting conditions are not symmetrical, then the system will ineluctably arrive at one of the stationary states, as fixed by the actual asymmetry in the starting conditions. Hence, *information is received.*

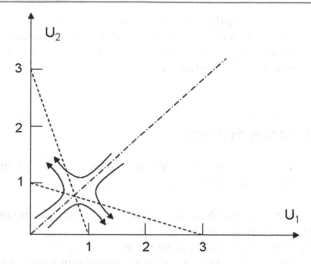

Fig. 7.3 Phase portrait of the system represented by Eq. (7.28) with $a = 1/3$. The main isoclines (cf. 7.28) are $u_1 = 0$ and $u_2 = 1 - au_1$ ("vertical", determined from $F_1 = u_1 - u_1u_2 - au_1^2 = 0$ with $\Delta u_1 = 0$), and $u_2 = 0$ and $u_1 = 1 - au_2$ ("horizontal", determined from $F_2 = u_2 - u_1u_2 - au_2^2 = 0$ with $\Delta u_2 = 0$), shown by *dashed lines*. The system has four stationary states: at $u_1 = u_2 = 0$, unstable, $\lambda_1 = \lambda_2 = +1$; at $u_1 = u_2 = 1/(1+a)$, unstable (saddle point), $\lambda_1 = -1$, $\lambda_2 = (1-a)/(1+a) > 0$; at $u_1 = 1/a$, $u_2 = 0$, stable, $\lambda < 0$; and at $u_2 = 1/a$, $u_1 = 0$, stable, $\lambda < 0$. The separatrix (separating the basins of attraction) is shown by the *dashed-dotted line* (after Chernavsky)

On the other hand, if the starting conditions are symmetrical (the system starts out on the separatrix), the subsequent evolution is not predetermined and the ultimate choice of stationary state occurs by chance. Hence, *information is generated*.[18]

7.3.3 Habituation

Empirical observation of many systems over time reveals that their responses to regularly repeated stimuli over time tend to decrease. This is called habituation or, especially when observed in a living system, fatigue. At the first sight this might seem paradoxical: it may be supposed that most real systems, of which the example in Sect. 7.3.1 is a simple illustration, are multistable and, hence, should potentially display considerable variety of behaviour. The explanation is that no matter how rich a system may be in states of equilibrium, after a time it will typically be found to be in a single basin of attraction.[19] Although both an initial endowment of potential variety of behaviour and ultimate stability seem like very necessary attributes for a

---

[18]Cf. the discussion in Chap. 2.
[19]See Ashby (1958) for a proof.

cell whose fate is to be a highly differentiated member of an organ, in other cases (such as an organism considered as a whole) it may be a great handicap. A random extraneous disturbance (i.e., noise) of sufficient amplitude may suffice to place the system in a different basin (dehabituation).

7.4   Evolutionary Systems

Equilibrium models, which are traditionally often used to model systems, are characterized by the following assumptions:

1. Entities of a given type are identical, or their characteristics are normally distributed around a well-defined mean
2. Microscopic events occur at their average rate
3. The system will move rapidly to a stationary (equilibrium) state (this movement is enhanced if all agents are assumed to perfectly anticipate what the others will do).

Hence, only simultaneous, not dynamical, equations need be considered, and the effect of any change can be evaluated by comparing the stationary states before and after the change.

The next level in sophistication is reached by abandoning assumption 3. Now, several stationary states may be possible, including cyclical and chaotic ones (strange attractors).

If assumption 2 is abandoned, nonaverage fluctuations are permitted, and the behaviour becomes much richer. In particular, external noise may allow the system to cross separatrices. The system is then enabled to adopt new régimes of behaviour, exploring regions of phase space inaccessible to the lower-level systems,[20] which can be seen as a kind of collective adaptive response (requiring noise) to changing external conditions.

The fourth and most sophisticated level is achieved by abandoning the remaining assumption, 1. Local dynamics cause the microdiversity of the entities themselves to change. Certain attributes may be selected by the system and others may disappear. These systems are called evolutionary. Their structures reorganize, and the equations themselves may change. Most natural systems seem to belong to this category. Rational prediction of their future is extremely difficult.

---

[20]This type of behaviour is sometimes called "self-organization"; cf. Érdi and Barna (1984).

References

Albert R, Barabási A-L (2002) Statistical mechanics of complex networks. Rev Mod Phys 71:47–97

Ashby WR (1956) An introduction to cybernetics. Chapman & Hall, London

Ashby WR (1958) The mechanism of habituation. In: Mechanization of thought processes, Proc NPL symposium. HMSO, London, pp 93–118

von Bertalanffy L (1993) Théorie générale des systèmes. Dunod, Paris

Érdi P, Barna Gy (1984) Self-organizing mechanism for the formation of ordered neural mappings. Biol Cybern 51:93–101

Fisher RA (1951) The design of experiments, 6th edn. Oliver & Boyd, Edinburgh

Galam S, Mauger A (1997) Universal formulas for percolation thresholds. Phys Rev E 53, 2177–2181; (ibid. 55:1230–1231)

Gardner MR, Ashby WR (1970) Connectance of large dynamic (cybernetic) systems: critical values for stability. Nature (Lond) 228:784

Huberman BA, Hogg T (1986) Complexity and adaptation. Physica D 22:376–384

Jongeling TB (1996) Self-organization and competition in evolution: a conceptual problem in the use of fitness landscapes. J Theor Biol 178:369–373

Kauffman SA (1984) Emergent properties in random complex automata. Physica D 10:145–156

Luthi PO, Preiss A, Chopard B, Ramsden JJ (1998) A cellular automaton model for neurogenesis in Drosophila. Physica D 118:151–160

Raine DJ, Norris VJ (2002) Network structure of metabolic pathways. J Biol Phys Chem 1:89–94

Robinson JC (1998) All possible chaotic dynamics can be approximated in three dimensions. Nonlinearity 11:529–545

Watts DJ, Strogatz SH (1998) Collective dynamics of small-world networks. Nature 393:440–442

Wolfram S (1983) Statistical mechanics of cellular automata. Rev Mod Phys 55:601–644

Algorithms

8

The concept of algorithm is of central importance, especially for arithmetic, and even more particularly for operations carried out by mathematical machines such as digital computers. An algorithm is defined as a process of solving problems based on repeatedly carrying out a strictly defined procedure. A classical example is the Euclidean algorithm for finding the greatest common divisor of two natural numbers a and b.

Example. Suppose $a > b$; divide a by b to yield either the quotient q_1 or the remainder r_2 (if b does not divide a), that is,

$$a = bq_1 + r_2, \quad 0 < r_2 < b. \tag{8.1}$$

Then if $r_2 \neq 0$, divide b by r_2:

$$b = r_2 q_2 + r_3, \quad 0 < r_3 < r_2, \tag{8.2}$$

and continue by dividing r_2 by r_3 until the remainder ineluctably becomes zero. Writing

$$r_{n-2} = r_{n-1} q_{n-1} + r_n, \tag{8.3}$$

$$r_{n-1} = r_n q_n, \tag{8.4}$$

then it is clear that r_n is the greatest common divisor of a and b.

By way of explanation, note that if two integers l and m have a common divisor d, then for any integers h and k, the number $hl + km$ will also be divisible by d. Denoting the greatest common divisor of a and b by δ, from Eq. (8.1) it is clear that δ is a divisor of r_2, from Eq. (8.2) it is also a divisor of r_3, and from Eq. (8.3) it is also a divisor of r_n, which is itself a common divisor of a and b, since from these equations it also follows that r_n divides r_{n-1}, r_{n-2}, and so forth. Thus δ is identical with r_n, and the problem is solved. This example is a well-defined procedure that leads automatically to the desired result.

An operation frequently required in bioinformatics is sorting a collection of items (an array of elements), implying arranging them in increasing (or decreasing) order. The so-called bubble sort (elements "float" to the top of the array) is considered to be the simplest algorithm. Each element is compared pairwise to each other. If a pair is

© Springer-Verlag London 2015
J. Ramsden, *Bioinformatics*, Computational Biology 21,
DOI 10.1007/978-1-4471-6702-0_8

found to be in the incorrect order, the two elements are interchanged. The algorithm is based on two DO-loops (repeat the instructions within the loop for a preset number of times, or until some condition is fulfilled), one nested inside the other. The outer loop runs from 1 to 1 − (the length of the array), and the inner loop runs from 1 + (a counter of the outer loop) up to the length of the array.

This sort algorithm is not particularly efficient, in the sense that algorithms requiring fewer instructions to accomplish the same task are available. Often these more efficient algorithms take more time to program, however. For example, the fast Fourier transform does indeed require significantly fewer instructions than the ordinary Fourier transform, but nowadays, with the almost universal availability of personal computers, provided the dataset being transformed is not too large, the extra work of programming might not be worth the bother. Most personal computers are switched off at night when they could actually be calculating. The Intel Pentium chip, introduced around 1996, can carry out 100 million instructions per second (MIPS); this is 5 times faster than the 486 chip, introduced around 1992, and 100 times more than the mainframe DEC VAX 780, introduced in 1980, and for more than a decade the workhorse of many computing centres. The DEC PDP1, again very widely encountered in its day, and introduced around 1960, carries out 0.1 MIPS. IBM's Deep Blue, also introduced in 1996, can accomplish 10^6 MIPS. Current processors such as the top of the range Intel Core i7 4770k operate at nearly 130,000 MIPS, while the Blue Gene Q supercomputer with thousands of processors is capable of 20 petaFLOPS (2×10^{16} floating point operations per second which, depending on architecture, can be many tens of MIPS). When computing jobs were processed batchwise on a mainframe device there was, of course, strong pressure to achieve operations such as sorting and matching with as few instructions as possible; but when the ubiquitous personal computer has 100 times more processing power than a VAX 780, the effort of achieving it may be considered superfluous by all who are not professional programmers.

Problem. Write a program to implement the bubble sort algorithm in a high-level computer language.

Problem. Write an algorithm for searching for all occurrences of a particular word (a substring) in a string and returning the distance of each occurrence from the start of the string.

8.1   Evolutionary Computing

Evolutionary computation (EC) is typically fairly informally defined as the field of computational systems that get inspiration and ideas from natural (Darwinian) evolution (cf. Sects. 10.9 and 10.9.2). One of the most important types of evolutionary computation is the genetic algorithm (GA), which is a type of search and optimization based on the mechanisms—albeit rather simplified—of genetics and natural selection. Each candidate solution is encoded as a numerical string, usually

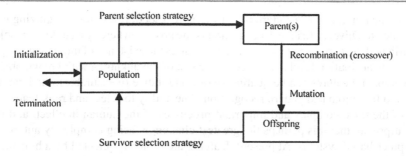

Fig. 8.1 An example of a genetic algorithm. One complete cycle constitutes one generation. Survival selection strategy determines which offspring, and which parents, are allowed to pass through to the next generation and which of those are allowed to become parents in the next cycle

binary (of course, unless an analog computer is used, ultimately even a real-valued string is encoded in binary form for processing on a digital computer). This string is called the chromosome. A large number of candidate solutions are then "mated": in other words, pairs of parents are selected (typically randomly) and the two chromosomes are mixed using operations inspired by those taking place in living cells (cf. Sect. 10.6), such as recombination (crossover). Random mutations to individual chromosomes are usually also allowed. The offspring are then evaluated according to some appropriate fitness criterion and mapped onto a numerical scale. Offspring with fitness below the threshold are eliminated. In some genetic algorithms, only the surviving offspring pass on to the next generation and all parents die; in others, the parents are also evaluated and retained if their fitness exceeds the threshold. The survivors then undergo another round of randomization and evaluation, and so on (Fig. 8.1). The cycles continue until a satisfactory solution is reached. The technique is particularly valuable for multiobjective optimization (MOO). Currently, there is much activity in the field, albeit dominated by heuristic developments. It is clear that there are many degrees of freedom available, and it would be impracticable in most cases to systematically investigate them all. A very promising trend is to allow more flexibility in the individual steps; ultimately, the algorithm should be able to develop itself under the constraint of some externally imposed fitness criterion. There is also a trend to more intensively apply some of the more recent discoveries in molecular biology to evolutionary computation, especially those regarding the epigenetic features known to control genome organization (see Sects. 10.8.2 and 10.8.3).

8.2 Pattern Recognition

Ultimately, pattern is a psychological concept: A set of objects fulfilling conditions of unity and integrity, according to which groups of objects with some common feature(s) are denoted and perceived (i.e., distinguished from other objects in their environment) by a *human being*. Pattern is therefore synonymous with class, category, or set. The remark that "a pattern is equivalent to a set of rules[1] for recognizing it" is attributed to Oliver Selfridge. Recognition is the process whereby an unknown object is attributed to a certain pattern (and hence requires the existence of more than one pattern). The attribution level involves comparison of the unknown with known objects (prototypes). Features can be qualitative or quantitative (measurable); the latter are required for automated pattern recognition. The ability to select and rank features is one of the most complex and important processes of the human intellect, and it is not surprising that it is perhaps the greatest challenge facing completely automated computer-based systems. At present, features are typically selected by a human.

The basic steps of pattern recognition are as follows:

1. Choice of the initial feature set. The number of features determines the dimensionality of feature space.
2. Measurement of the chosen features of a prototype.
3. Preparation (elimination of excess information—noise),[2] resulting in a somewhat standardized description (a prototype), which is then used to construct the training set.
4. Construction of the decision-making rule.
5. Comparison of any (typically prepared) unknown object with a prototype; with the help of a quantitative resemblance measure, a decision is made whether the unknown object belongs to the pattern.

Pattern recognition is thus seen to be a supervised (i.e., undertaken with a teacher) learning process. Learning implies that the decision-making rule is modified by experience. The process of pattern recognition is typically computationally heavy; thus, in this field there is a strong motivation for finding algorithms that are very efficient.

The discernment of clumps or clusters of objects according to the features chosen to represent them transcends the recognition of patterns in the sense of noting the similarity of a known object to an unknown object. Where data are simply analysed and clusters are found, this is pattern discovery and is dealt with in the next section.

8.3   Botryology

The term "botryology", apparently coined by Good (1962), was introduced at a time when the task of finding clusters was generally focused on objects arranged in ordinary (Euclidean) space (e.g., stars clustered into galaxies). It signifies a more general approach to finding clusters or clumps, concerned with logical and

---

[1]I.e., an algorithm.
[2]For example, imagine a typical time-varying signal such as the output of a microphone. This can be converted to a square wave of uniform amplitude and varying period.

qualitative relationships, chosen for their relevance to the matter in hand, rather than with ordinary distance (or a metric satisfying a triangle inequality; see below). Possibly relevance could be defined according to success in finding clusters or clumps, hence permitting iterative refinement of the definition.

Since then the notion of clustering has anyway been somewhat generalized, and typically now includes any process whereby relevance can lead to a numerical attribute (e.g., the conditional probability of use of an object). The objects are nodes on a graph (Sect. 7.2), and the links between them (edges) give the relevance. Thus, an element a_{ij} of the adjacency matrix A gives the relevance of i to j. This may not be the same as a_{ji}, giving the relevance of j to i; hence, the graph is a directed one. On the other hand, association factors such as $P(ij)/P(i)P(j)$ (the probability of the joint occurrences divided by the product of the probabilities of the separate occurrences) are symmetrical. The degree of clumpiness of a group of nodes could then be given by summing the elements of the adjacency matrix of the group and dividing by the number of elements in the group; a clump could be considered as complete if the addition of an extra node would bring the clumpiness below some threshold.

Possibly it is useful to use the term "clustering" for the formal process (which can be carried out on a computer) described in Sect. 8.3.1 and the term "clumping" for a more general process (of which clustering would be a subset), for which formal definitions might not always be available.

It is possible to conceive a highly automated mode of scientific investigation, in which every object in the universe would be parametrized (by which is meant that a numerical value is assigned to every attribute). In order to investigate something more specifically, the researcher would select the relevant collection of objects (e.g., "furry mammals") and apply some kind of dimensional reduction to the dataset (if the attributes were chosen from some vast standardized set, many would, of course, have values of zero for a particular collection), preferably down to two or three,[3] after which a clustering algorithm would be applied.[4]

8.3.1 Clustering

Whereas supervised pattern recognition (i.e., with a teacher) corresponds to the most familiar kind of pattern recognition carried out by human beings throughout their waking hours (in other words, the comparison of unknown objects with known prototypes), of more current interest in bioinformatics is the unsupervised discernment of patterns in, for example, gene and genome sequences, especially since the pro-

---

[3]E.g., using principal component analysis (PCA) (q.v.).
[4]See Gorban et al. (2005) for an example.

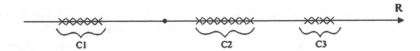

Fig. 8.2 Each object is represented by a cross corresponding to its value of the chosen feature on the real line $\mathbb{R}$. Clusters *C1*, *C2*, and *C3* are easily identifiable. The spot on the line represents a possible value that could be used to divide the set dichotomously

portion of unknown material in genomes is still overwhelming.[5] A very powerful methodology for achieving that is to examine whether the data resulting from some operations carried out on a DNA sequence (for example) can be arranged in such a way that structure appears, namely that groups of data points constituting a subset of the entire dataset are clumped together to form two or more distinct entities.

The clustering process is defined as the partition of a set of objects by some features into disjoint subsets, and each subset in which objects are united by some features is called the cluster. If no relation between the objects is known, it is impossible to construct clusters.

The simplest case of clustering arises when only one feature exists; each object under consideration either has the feature or does not, in which case the maximum number of clusters is two and, if it happens that all the objects have that feature, then there will be only one cluster. Another simple case arises if values from the real-number line can be attributed to the feature (Fig. 8.2). This is easily generalized to two or more dimensions, the number of dimensions being equal to the number of chosen features.

If the set of objects is large and many features have been chosen, it is necessary to have algorithms for clustering that enable it to be carried out automatically on the computer. Many such algorithms are known; a few of them are briefly described below. It is assumed that there is a set $\{X\}$ of objects (X_i, etc.) in N-dimensional feature space. For ease of representation, we will tacitly consider $N = 2$.

Hyperspheres A circle of radius r is drawn around an arbitrarily chosen object. Objects within the circle form the first subcluster. New circles are now drawn with their centres at these other objects, which encompass yet more objects, around which new circles are again drawn, and so forth until no new objects are added. If all of the objects in the set are now included, the process has failed. If, on the other hand, objects remain, then one of those remaining objects is arbitrarily chosen and the process is repeated.

The radius r must fall between the minimum and maximum distances between the objects. The larger its value, the fewer will be the number of clusters. Possibly other criteria are needed to select the most appropriate value (e.g., from some prior

---

[5]We also have the intermediate process of semisupervised learning, which deals with the problem of combining small amounts of labelled data with large amounts of unlabelled data—the classic paper is Zhu et al. (2003).

estimation of the likely number of clusters). The method of dynamic kernels is analogous to hypersphere clustering.

The K-means Method This method originated from the so-called ISODATA (iterative self-organizing data analysis) technique. The centres of K clusters are chosen simultaneously. Denoting the centre of the kth cluster by Z_k, $k = \overline{1, K}$, then for the process of cluster formation, in particular for the incorporation of any object X into cluster C_k, we have

$$X \in C_k \text{if } \rho(X; Z_k) \leq \rho(X; Z_i) , \tag{8.5}$$

where $k = \overline{1, K}$, $i \neq k$. In the next step, new centres of gravity for the K subclusters are computed. In the step l, for each new dividing D_l the functional $F(D_l)$ is computed by the expression

$$F(D_l) = \sum_{X \in C_{kl}} (X - Z_{kl})^2 . \tag{8.6}$$

The optimal division is that for which the function F takes its minimal value. The process of dividing goes on until for the centres of the next two steps the condition

$$Z_{k,l+1} = Z_{kl} \tag{8.7}$$

is satisfied. The effectiveness of this algorithm depends on the chosen value of K, the selection of the initial clustering centres, and the actual location of the points in feature space corresponding to the objects, which together constitute a significant weakness of this method.

Distance Metrics The calculation of a distance between any two objects is fundamental to clustering. In Euclidean space, the operation is intuitively straightforward, especially when the positions of each object in space are represented using Cartesian coordinates. Thus, in one dimension, the distance between two objects at positions x_1 and x_2 is simply their difference, $|x_1 - x_2|$. The procedure is generalized to higher dimensions using familiar knowledge of coordinate geometry (Pythagoras' theorem); thus, for two orthogonal axes x and y, the distance is $\sqrt{(x_1 - x_2)^2 + (y_1 - y_2)^2}$. The space must be chosen according to relevance. Thus, a collection of trees might be characterized by height and the mean rate of photosynthesis per unit area of leaf. Each member of the collection (set) would correspond to a point in this space. An explicit procedure must be provided for assigning numerical values to these two parameters. *Ex hypothesi*, they are considered to be independent; hence, the axes are orthogonal. Especially when the number of dimensions of the chosen space is high, it is convenient to reduce it to two, because of the inescapable convenience of representing data as a picture in two dimensions. For this purpose, principal component analysis, described in the next Sect. 8.3.2, is a useful method.

8.3.2   Principal Component and Linear Discriminant Analyses

The underlying concept of principal component analysis (PCA) is that the higher the variance of a feature, the more information that feature carries. PCA, therefore, linearly transforms a dataset in order to maximize the retained variance while minimizing the number of dimensions used to represent the data, which are projected onto the lower- (most usefully two-) dimensional space.

The optimal approximation (in the sense of minimizing the least-squares error) of a D-dimensional random vector $\mathbf{x} \in \mathbb{R}^D$ by a linear combination of $D' < D$ independent vectors is achieved by projecting $\mathbf{x}$ onto the eigenvectors (called the principal axes of the data) corresponding to the largest eigenvalues of the covariance (or scatter) matrix of the data represented by $\mathbf{x}$. The projections are called the principle components. Typically, it is found that one, two, or three principal axes account for the overwhelming proportion of the variance; the sought-for reduction of dimensionality is then achieved by discarding all of the other principal axes.

The weakness of PCA is that there is no guarantee that any clusters (classes) that may be present in the original data are better separated under the transformation. This problem is addressed by linear discriminant analysis (LDA), in which a transformation of $\mathbf{x}$ is sought that maximizes intercluster distances (e.g., the variance between classes) and minimizes intracluster distances (e.g., the variance within classes).

8.3.3   Wavelets

Most readers will be familiar with the representation of arbitrary functions using Fourier series, namely an infinite sum of sines and cosines (called Fourier basis functions).[6] This work engendered frequency analysis. A Fourier expansion transforms a function from the time domain into the frequency domain. It is especially appropriate for a periodic function (i.e., one that is localized in frequency), but is cumbersome for functions that tend to be localized in time. Wavelets, as the name suggests, integrate to zero and are well localized. They enable complex functions to be analysed according to scale; as Graps (1995) points out, they enable one to see "both the forest and the trees". They are particularly well suited for representing functions with sharp discontinuities, and they embody what might be called scale analysis.

The starting point is to adopt a wavelength prototype function (the analysing or mother wavelet) $\Phi(x)$. Temporal analysis uses a contracted, high-frequency version of the prototype, and frequency analysis uses a dilated, low-frequency version. The wavelet basis is

$$\Phi_{s,l}(x) = 2^{-s/2}\Phi(2^{-s}x - l) \,, \tag{8.8}$$

---

[6]Fourier's assertion was that any 2π-periodic function $f(x) = a_0 + \sum_{k=1}^{\infty}(a_k \cos kx + b_k \sin kx)$. The coefficients are defined as $a_0 = (2\pi)^{-1}\int_0^{2\pi} f(x)\,\mathrm{d}x$, $a_k = \pi^{-1}\int_0^{2\pi} f(x)\cos(kx)\,\mathrm{d}x$, and $b_k = \pi^{-1}\int_0^{2\pi} f(x)\sin(kx)\,\mathrm{d}x$.

where the variables s (wavelet width) and l (wavelet location) are integers that scale and dilate Φ to generate (self-similar) wavelet families. If different resolutions are required, a scaling function $W(x)$, defined as

$$W(x) = \sum_{k=-1}^{N-1} (-1)^k c_{k+1} \Phi(2x + k) , \tag{8.9}$$

is used, where the c_k are the wavelet coefficients, which must satisfy the constraints $\sum_{k=0}^{N-1} c_k = 2$ and $\sum_{k=0}^{N-1} c_k c_{k+2l} = 2\delta_{l,0}$, where δ is the delta function.

The wavelet transform is the convolution of signal and basis functions:

$$F(s, l) = \int f(x) \Phi_{s,l}^*(x) \, dx \tag{8.10}$$

where Φ^* is the complex conjugate of Φ. Often, the data can be adequately represented as a linear combination of wavelet functions, and their coefficients are all that is required for carrying out further operations on the data.

8.4 Multidimensional Scaling and Seriation

Multidimensional scaling[7] (MDS) provides a means of estimating the contents of a vector space of data from a given minimum set of input data. The N objects or vectors under consideration are characterized by a quantity M of parameters common to all the objects. In estimating the relative values of the parameters for each object, the original vector space may be reconstructed from $N(N/2 - 1)$ pieces of data; that is, $N \times M$ elements of data are thus recovered. An important application of MDS is the reconstruction of an original M-dimensional vector space from one-dimensional distance data between vectors of the space.

Known Data Consider an M-dimensional vector space containing N vectors. The vectors may be considered as N objects containing M possible parameters or unit vectors. The objects are then characterized by the scaling of the unit vectors. Suppose that the only known information concerning the object structure is a distance measure between each of the N objects, given by a symmetric $N \times N$ matrix.

Estimating Data For each vector, an M-dimensional initial estimated vector is formed from a random seed and then propagated iteratively. The propagation is determined such that each iteration minimizes a stress function (i.e., a normalized measure of the distance between the distance matrix estimate and the given distance matrix vectors). Iteration continues until a defined minimum of the stress function is found; a representation of the original M-dimensional space of N vectors may then be displayed from the estimated vectors.

---

[7] See Kruskal (1964).

Theory Define the M-dimensional vector space of N objects by the vectors

$$\mathbf{x}_i = \sum_{\mu=1}^{M} b_{i\mu}\hat{y}_\mu, \tag{8.11}$$

where $\hat{y}_\mu$ are the unit vectors of the space. The Euclidean distances between these vectors are then given by the $N \times N$ distance matrix

$$E_{ij} = [(\mathbf{x}_i - \mathbf{x}_j)^2]^{1/2}. \tag{8.12}$$

If only this matrix is known and not the underlying vectors, then an estimated distance matrix may be defined:

$$\widetilde{E}_{ij} = [(\widetilde{\mathbf{x}}_i - \widetilde{\mathbf{x}}_j)^2]^{1/2}. \tag{8.13}$$

The estimated vectors may be formed as

$$\widetilde{\mathbf{x}}_i = \sum_{\mu=1}^{\widetilde{M}} a_{i\mu}\hat{y}_\mu, \tag{8.14}$$

where

$$a_{i\mu} = a_{0i\mu} + z_{i\mu} \tag{8.15}$$

and $a_{0i\mu}$ are initial values selected at random and $z_{i\mu}$ are used to propagate the vector through iteration.

The stress function S is a normalized measure of the distance between the distance matrix estimate and the given distance matrix vectors:

$$S^2 = \frac{\sum_{i,j=1}^{N,N}[\widetilde{E}_{ij} - E_{ij}]^2}{\sum_{i,j=1}^{N,N} E_{ij}}. \tag{8.16}$$

This may be minimized by

$$\frac{\partial S^2}{\partial z_{k\mu}} = 0, \tag{8.17}$$

but E_{ij} is constant and given by

$$B = \sum_{i,j=1}^{N,N} E_{ij}, \tag{8.18}$$

so that

$$\frac{\partial S^2}{\partial z_{k\mu}} = 2B^{-1} \sum_{i,j=1}^{N,N} [\widetilde{E}_{ij} - E_{ij}] \frac{\partial \widetilde{E}_{ij}}{\partial z_{k\mu}}. \tag{8.19}$$

Using Eq. (8.14) gives

$$\frac{\partial \widetilde{E}_{ij}}{\partial z_{k\mu}} = \widetilde{E}_{ij}^{-1} \sum_{\nu=1}^{\widetilde{M}} [a_{i\nu} - a_{j\nu}][\delta_{ik}\delta_{\nu\mu} - \delta_{jk}\delta_{\nu\mu}], \tag{8.20}$$

where the Kronecker delta δ_{ik}, as usual, equals 1 for $i = k$ and 0 for $i \neq j$. Then, after some algebra,

$$\frac{\partial S^2}{\partial z_{k\mu}} = 4B^{-1} \sum_{j=1}^{N} [\widetilde{E}_{kj} - E_{kj}]\widetilde{E}_{kj}^{-1}[a_{k\mu} - a_{j\mu}]. \tag{8.21}$$

Hence, by integration, the estimated vectors are given by

$$\widehat{z}_{k\mu} = z_{k\mu} + \alpha\frac{\partial S^2}{\partial z_{k\mu}}, \tag{8.22}$$

where $\widehat{z}_{k\mu}$ is the next iteration, and minimizing the stress function provides the scale and direction for the propagation, and α provides the iteration increment, typically fixed as N^{-3}. Iteration continues until the stress function reaches zero or some lower threshold. Note that the value of $\widetilde{M}$ used to reconstruct the vector space need not be the same as the original space dimension M.

An important application of MDS is to seriation—the correct ordering of an assembly of objects along one dimension, given merely the presence or absence a certain number of features in each object.[8] These data are arranged in a Boolean incidence matrix, with the rows corresponding to the objects and the columns to the features, a "1" corresponding to the presence of a feature in an object. The characteristic pattern to be expected is that in every column, the 1s are clumped together, or, if there are multiple representations of features in the objects, in every column their number increases to a maximum and then decreases. Evidently, this can be achieved by appropriate rearrangement of the order of the rows. All of the relevant information is contained in the similarity matrix (in the sense of similar to the serial ordering), in which the element (i, j) is the number of features common to the ith and jth objects.

8.5   Visualization

It seems almost impossible to overestimate the power of visualization, as a mode of knowledge representation, to influence the interpretation of data.[9] In this regard, supremacy belongs to Cartesian coordinates, perhaps the most important mathematical invention of all time. Two-dimensional representations that can be drawn on paper (or viewed on a screen) are particularly significant. As already mentioned, one of the main motivations of PCA (Sect. 8.3.2) is to enable a complex dataset to be represented on paper. This applies equally well to dynamical representations of evolving

---

[8]This was famously applied by Kendall (1970) to the problem of chronology of early Egyptian tombs found at a certain site. The features in that case are artisanal artefacts characteristic of a certain epoch found in the tombs.

[9]Cf. Sect. 22.2.

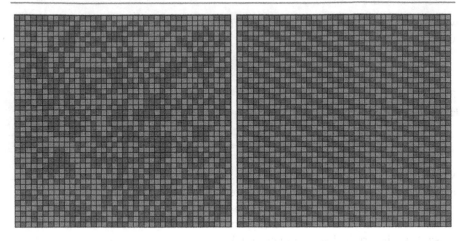

Fig. 8.3 The binary expansion of the first 1600 decimal digits (mod 2) of π (*left*) and 22/7 (*right*), represented as an array of *light* (0) and *dark* (1) *squares*, to be viewed left to right, top to bottom

systems, in which phase portraits (state diagrams in phase space; cf. Fig. 7.3) of a dynamical system such as a living cell can be very influential.

Another kind of visualization consists in generating images from binary expansions.[10] On paper, both the actual decimal digits of the irrational number π and those of the rational approximation 22/7 look random; when their binary expansions are drawn as rows of white (corresponding to 0) and black (corresponding to 1) squares, pattern (or its absence) is immediately discernible (Fig. 8.3).

More generally, visualization should be considered as part of the overall process of accumulating convincing evidence for the validity of a proposition. It should not, therefore, be merely an alternative to a written or verbal representation, but should transcend the limitations of those other types of representation.

References

Good IJ (1962) Botryological speculations. In: Good IJ (ed) The scientist speculates. Heinemann, London, pp 120–132

Gorban AN, Popova TG, Zinovyev A (2005) Codon usage trajectories and 7-cluster structure of 143 complete bacterial genomic sequences. Phys A 353:365–387

Graps A (1995) An introduction to wavelets. IEEE Comput Sci Eng 2:50–61

Kendall DG (1970) A mathematical approach to seriation. Phil Trans R Soc A 269:125–135

---

[10]It is said that Leibniz was the first to raise this possibility in a letter to one of the Bernoulli brothers, in which he wondered whether it might be possible to discern a pattern in the binary expansion of π.

Kruskal JB (1964) Multidimensional scaling by optimizing goodness of fit to a nonmetric hypothesis. Psychometrika 29:1–27

Zhu X, Ghahramani Z, Lafferty J (2003) Semi-supervised learning using Gaussian fields and harmonic functions. In: Proceedings of 20th international conference on machine learning (ICML-2003), Washington, DC

Part II
Biology

Introduction to Part II

9

The primary purpose of this and the following two chapters is to give an overview of living systems, especially directed at the bioinformatician who has previously dealt purely with the computational aspects of the subject.

Whenever confronting the totality of biology, it is clear that one may approach it at various levels, such as molecular, cellular, organismal, populational and ecological. Traditionally, these levels have been accorded official status by naming academic departments after them. Just as we saw with the levels of information (technical, semantic, effective), however, one quickly distorts a vision reflecting reality by insisting on the independence of the levels. For example, it not possible to understand how populations of organisms evolve without considering what is happening to their DNA molecules. When reading the two following chapters, this interdependence should constantly be borne in mind.

Problem. Attempt to provide a definition of life. Find exceptions.

9.1   Genotype, Phenotype, and Species

The basic unit of life is the organism. The phenotype may be defined as the organism interacting with its environment. The genotype may be defined as the set of instructions for the self-reproduction of the organism and is supposed to be barely influenced by external conditions.

Those adhering to the primacy of the genome will nevertheless concede that sending the complete gene sequence of an organism to an alien civilization will not allow the reconstruction of the organism (i.e., the creation of a living version of it—i.e. its phenotype). Many things, including the principles of chemical catalysis necessary for the genetic instructions to be read and processed, are not represented and are not even implicit in the nucleic acid sequence. In fact, the phenotype is a composite of explicit and implicit meaning, the latter being context-dependent.

© Springer-Verlag London 2015
J. Ramsden, *Bioinformatics*, Computational Biology 21,
DOI 10.1007/978-1-4471-6702-0_9

Table 9.1 Approximate equivalents of contrasting concepts

Genotype	Phenotype
Genetics	Epigenetics
Genome	Proteome
Nature	Nurture
Gene	Environment
Necessity	Chance (or freedom)
K	I (from Eq. (2.13))
Explicit	Implicit
Semantics	Syntax
...	...

It must also be kept in mind that the processes of natural selection, to be considered more deeply in Chap. 10, operate on the phenotype, yet the vehicle for their persistence is the genotype.

Organisms are commonly characterized as species. Despite the pervasive use of the term in biology, no entirely satisfactory definition of "species" exists. "Reproductive isolation" is probably one of the better operational definitions, but it can only apply under carefully circumscribed conditions. Geographical as well as genetic factors play a rôle, and epigenetic factors are even more important. In any human settlement of at least moderate size, there are almost certainly groups of inhabitants with no social contact with other groups. Hence, these groups are as effectively reproductively isolated from each other, because of behavioural patterns, as if they were living on different continents, and if we apply our definition, we are forced to assert that the groups belong to different species (even though are all taxonomically classified as *Homo sapiens*).

The concept of reproductive isolation is of little use when species reproduce asexually (such as bacteria); in this case, a criterion based on the possibility of significant exchange of genetic material with other organisms may have to be used.[1]

Another difficulty in defining "species" in terms of associating them with autonomously reproducing DNA is that not only are there well-defined organisms such as coral or lichen in which two "species" are actually living together in inseparable symbiosis, but we ourselves host about 10^{14} unicellular organisms, mostly bacteria, which comfortably outnumber the 10^{13} or so of our own cells.

A very striking characteristic of living organisms is that they are able to maintain their being in changing surroundings. It is doubtful whether any artificial machine can survive over as wide a range of conditions as man, for example. "Survival" means that the essential variables of the organism are maintained within certain limits. This maintenance (homeostasis) requires regulation of the vital processes. We shall consider regulation more formally in the following Sect. 9.2.

---

[1] See also Sect. 22.3.

Table 9.1 puts some of the terms encountered into a kind of correspondence. It is not a table of synonyms.

Problem. Discuss and extend Table 9.1. Find descriptive headings for the columns.

9.2 Adaptation

It can well be stated that *adaptation* is perhaps the most characteristic feature of life. The process of adaptation, so ubiquitous in nature, has been formalized by Sommerhoff (1950). The "disturbance" (cf. Sect. 9.4) presented at epoch t_0 is denoted the *coenetic variable* D_{t_0}, the "hardware" (approximately equivalent to T in Sect. 9.4) is the environmental circumstance E_{t_1}, the "regulator" (approximately equivalent to R in Sect. 9.4) is the response R_{t_1} directively correlated with E_{t_1}, both R and E taking place at a particular subsequent epoch t_1, and the "essential variables" (cf. Sect. 9.4) constitute the focal condition or goal G of the organism that reaches its consummation at the still later epoch t_2.

The usual notion of adaptedness, as applied to biological systems, implies no more than appropriateness. In other words, the statement that an (organic) response R is adapted to the environmental circumstances E from the viewpoint of some future state of affairs G (toward the realization of which it is considered to be directed) implies that the response is appropriate and, hence, also effective in bringing about the actual (or at least the probable) occurrence of G. However, although this "definition" of adaptedness is easy to state, it is not only trivial in meaning but is also fraught with difficulties. For one thing, it does not allow us to prefer the statement "the fish is adapted to the aquarium in which it survives" to "the aquarium is adapted to the fish it contains". Another difficulty is presented by that numerous category of accidental activity. Many accidental occurrences (including random mutations of DNA) are highly effective in bringing about a certain response but could hardly be called adapted; in the case of a random mutation, for example, adaptation could be said to have occurred only after it had become fixed in the population due to the advantages it conferred on the organism.

In Sommerhoff's formulation, adaptation (i.e., the statement that R_{t_1} is adapted to E_{t_1} with respect to G_{t_2}) means that if a changed disturbance D_{t_0} caused the occurrence of an alternative member of the set of Es (environmental circumstances), it would also have caused the occurrence of an alternative member of the set of Rs (appropriate responses) such that the goal G_{t_2} would still have been achieved. In other words, the response R is not only appropriate given the actual environmental circumstance E but would also have been appropriate had the initial disturbance D been different. It should be emphasized that E and R are epistemically independent variables (if they were dependent, then achievement of the goal would merely be a manifestation of physical stability). The disturbance D is called the coenetic variable, underlying the fact that it is a common causal determinant of both E and R. Directive correlation is this special relationship between E and R (Fig. 9.1). Its existence renders the goal

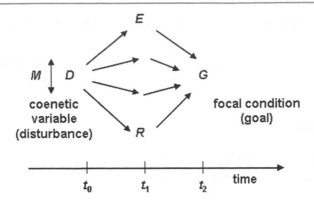

Fig. 9.1 Directive correlation (after Sommerhoff 1950). The *arrows* indicate causal connexions. In this drawing, four correlated variables (E and R) are involved. See the text for explanation of the symbols

independent of D. Adaptation is thus a tetradic relationship among D (which may be a prior occurrence of E), E, R, and G. Furthermore, it is not necessary to restrict the coenetic variable to specific environmental stimuli that evoke an organic response; it can also be a general factor that determines the specific nature of an action. It may also be remarked that the general purpose of sensory organs is to establish those causal connexions that will enable environmental variables to become the coenetic variables of adapted organic behaviour.

The degree M of directive correlation can be defined as the range of variation of the coenetic variable over which directive correlation can be maintained, and the range N of directive correlation can be defined as the number of correlated (E and R) or coenetic (D) variables involved. The degree is especially important because it is related to the minimum probability that the goal will be achieved.

9.3   Timescales of Adaptation

One can identify three timescales: proximate (short term, often associated with behaviour)—such as immediate response to sudden danger (e.g., fleeing from a fire); ontogenic, or the abilities that accumulate over the lifetime of an individual (medium term, often associated with learning, or a pattern of behaviour); and phylogenetic, or the inheritable changed capacities associated with changes in the genome, which constitute evolution of a species (long term). Proximate adaptation may take place through the medium of reception of information (e.g., a toxin binding to a cell surface receptor) followed by appropriate gene expression (cf. Sect. 9.4), but in many animal responses there is no time even for this, but simply for muscular action. The mechanisms for phylogenetic adaptation, involving DNA mutations, are now similarly well established. It is only in recent years, however, that a considerable repertoire of molecular mechanisms for ontogenic adaptation has been discovered, including the establishment of gene methylation patterns that more or less permanently (unless there is a drastic change in circumstance) fix which genes are potentially expressible in a given cell. The vast accumulation of nongenic ("noncoding") DNA in most eukaryotes is no doubt of great value here, permitting the synthesis of small interfering RNAs that gradually build up a repertoire for modulating gene expression according to the particular circumstances of the individual cell.

This rather clear-cut structure of adaptive timescales is not readily applicable to prokaryotes. First, their genomes are extremely plastic and can acquire genetic material from the environment throughout the lifetime of the organism. Second, the meaning of "lifetime of an individual" is not so clear: When a bacterium divides, does it really create two equal offspring, simultaneously annihilating itself? Does it essentially bud off excreta in a less vital, perhaps almost moribund version of the parent, which thereby gains a new lease of life? Does it gather its vital forces and concentrate them in a fresh new organism, accepting senescence and death for itself?

9.3.1 The Rôle of Memory

The picturesque idea of human (and, as far as we know, other animals') memory as a vast warehouse of facts to be retrieved at will, closely analogous to the digital memories of modern computers, would appear to be very far from the truth. Man, in particular, appears to possess immense power of bringing past experience (including that of fellow members of the species, via written or other records) to bear on the present situation. In terms of the schemata of Figs. 9.1 and 9.2, this input should be included in the regulatory response R.

9.3.2 The Integrating Rôle of Directive Correlation

Although the ultimate goal of any organism is survival, the functions of most of the individual organs are very subordinate to that ultimate goal. The goal of a subordinate function may simply be the maintenance of the physiological conditions required to

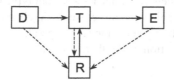

Fig. 9.2 Schematic diagram of a regulatory mechanism. The components are as follows: D, disturbance (from the environment); T, the hardware or the mechanism; R, the regulator; and E, the essential variables (output). *Arrows* represent communication channels along which information passes; those with *solid shafts* must exist, and those with *dashed shafts* may exist. See the text for further explanation

keep the coenetic variable of a higher function within its maximum permissible range of variation; in other words, there may be directive correlations of directive correlations carried on through many levels.

As the range of directive correlation increases, more and more causal connexions are required. This is particularly apparent when considering coordinated activities. An action such as running requires the coordination of many muscles; each one must take account of the others, and all have a common goal. n muscles may therefore require as many as $n^2 + n$ physical interconnexions. If the muscles are physically distant from each other, the construction and maintenance of these interconnexions may represent a considerable burden; but if they are concentrated within a nervous centre, only n afferent and n efferent connexions are required, together with n more leading to the goal itself; physical economy in the total length of the connexions provides a natural explanation for the existence of nerve centres.

Clearly, directive correlation is practically synonymous with organic integration, bringing into connexion (through the objective property of directive correlation) what would otherwise be independent, disconnected entities.

9.4   Regulation

Regulation may be considered in abstract terms common to any mechanism, whether living or not. The formalism presented below will be explicitly made use of in Chap. 16 when considering signalling and regulatory pathways.

The essential elements of a regulatory system are shown in Fig. 9.2. The lines connecting the components indicate communication channels. The dotted lines indicate the paths along which the regulator can receive information about the disturbance. By way of illustration, consider the operation of a simple thermostatted water bath. T then represents the electric heater and the bath itself with a circulator. E represents the water temperature (measured with a thermometer) T, R represents the switch controlling the power supplied to the heater, and D represents the disturbances from the environment. A typical event is the immersion of a flask containing liquid at a temperature lower than that of the bath. Sophisticated baths may be able to sense the temperature and mass of the flask before it has been immersed (channel D $\rightarrow$ R), or at the moment of its placement (channel T $\rightarrow$ R), but typically the heater is switched on if the temperature falls below the target value T_0 (channel E $\rightarrow$ R). This is called regulation by error. Most living cells appear to operate according to this principle.

The canonical representation of the thermostat is

$$\downarrow \begin{matrix} a & b \\ a & a \end{matrix} , \tag{9.1}$$

where state a represents $T = T_0$ (within the allowed uncertainty) and state b represents $T < T_0$.

In the case of a bacterium, a may, for example, represent $[Hg^{2+}] = 0$ (square brackets denoting concentration) and b may represent $[Hg^{2+}] > 0$. D in Fig. 9.2

now corresponds to mercury ions in the environment of the cell, T corresponds to the proteins able to sense mercury ions and the gene expression machinery able to synthesize mercury reductase, and R corresponds to the transcription factor binding to the mercury reductase gene promoter sequence. In stochastic matrix representation, we have

$$
\begin{array}{c|cc}
\rightarrow & a & b \\
\hline
a & 1 & 0 \\
b & 1 & 0
\end{array} \ . \tag{9.2}
$$

More realistically, however, we might have

$$
\begin{array}{c|cc}
\rightarrow & a & b \\
\hline
a & 1.0 & 0.0 \\
b & 0.6 & 0.4
\end{array} \ , \tag{9.3}
$$

for example, since, for various reasons, the machinery may not work perfectly. Further sophistication may be incorporated by increasing the number of states; for example, a, b, c, and d corresponding respectively to $[\mathrm{Hg}^{2+}] = 0$, $1\,\mathrm{nM}$, $1\,\mu\mathrm{M}$, and $1\,\mathrm{mM}$ and above, with the corresponding matrix

$$
\begin{array}{c|cccc}
\rightarrow & a & b & c & d \\
\hline
a & 1.0 & 0.0 & 0.0 & 0.0 \\
b & 0.6 & 0.4 & 0.0 & 0.0 \\
c & 0.3 & 0.4 & 0.3 & 0.0 \\
d & 0.0 & 0.3 & 0.4 & 0.3
\end{array} \ .
$$

After several cycles, the machine will be completely in state a (cf. Sect. 6.2).

In the simplest cases, the error, or a quantity proportional to it, is sent back to the regulator but, more sophisticatedly, some function of the error—for example, its integral, or its derivative—could be fed back to R. The vast majority of industrial controllers use a combination of all three (and hence are referred to as PID controllers).

9.5 The Concept of Machine

"Machine" is used formally to describe the embodiment of a transformation (e.g., Eq. (9.1); cf. the automata in Sect. 7.11). The essential feature is that the internal state of the machine, together with the state of its surroundings, uniquely defines the next state to which it will go. A determinate machine is canonically represented by a closed, single-valued transformation Eqs. (9.1 and 9.2); a Markovian machine is indeterminate insofar as the transitions are governed by a stochastic matrix (e.g., Eq. 9.3); the determinate machine is clearly a special case of the more general Markovian machine.

If there are several possible transformations and a parameter governs which transformation shall be applied to the internal states of the machine, then we can speak of a machine with input, the input being the parameter. The machine with input is therefore a transducer (cf. Sect. 3.3).

A Markovian machine with input would be represented by a set of stochastic matrices together with a parameter to indicate which matrix is to be applied at any particular step. If these parameters are themselves controlled by a stochastic matrix, then we have a so-called hidden Markov model (Sect. 13.5.2).

9.6   The Architecture of Functional Systems

Almost any system is confronted with the problem that as its complexity increases, more and more channels of communication are required (cf. Sect. 9.3.2), with greater and greater information capacity, if every component of the system is to remain fully integrated. A very important way of coping with this problem is to organize systems hierarchically, such that the amount of information is distributed more or less uniformly across levels; by this means the information flow within and between levels remains manageable. One way of quantifying the degree of hierarchicality is to determine the distribution of path lengths between pairs of components; the closer it is to a power law distribution, the more hierarchical the system (cf. Chap. 7).

As the size of a system (as measured by the number of constituent components) increases, if every component had to be individually designed and fabricated, the burden of doing so would soon become overwhelming. In artificial systems, such as very large-scale integrated circuits, this problem is evaded by a combination of functional modularity and structural regularity. The latter is anything that reduces complexity, in the sense already discussed in Sect. 6.5 (e.g., the repetition of components). Thus, even the most sophisticated integrated circuits have essentially only two types of basic components, pMOS (p-type metal–oxide–semiconductor field-effect transistors) and nMOS (their n-type equivalents).

Functional modularity is the *structural localization of function*.[2] In other words, some function is separated into structural units ("modules"); these are able to carry out some information processing internally, which diminishes the amount of information that needs to flow between modules (cf. the rôle of nervous centres, Sect. 9.3.2). It may even arise that design principles developed for modules at one level in a hierarchy can be reused for modules at other levels. Functional modularity can also be quantified, provided that function and structure are quantifiable. The dependency of whole-system function on the components of an arbitrarily chosen piece of the system can then be measured. The less that dependency itself depends on components outside the chosen piece, the more the function of that piece is localized (i.e., the more modular it is). If the dependencies are represented as second derivatives of function with respect to pairs of parameters (the Hessian matrix of fitness), modules

---

[2]See Lipson (2007)

can be identified as those collections of parameters that are concentrated around the diagonal of the matrix.

Problem.   Quantify the regularity, modularity, and hierarchicality for a variety of artificial and natural systems.

9.7   Biological Complexity

It has long been a tenet of biology that there has been a gradual increase in phenotypic complexity during the history of life on Earth;[3] it is "what everybody knows",[4] although hard evidence has been remarkably difficult to come by, not least because of a lack of consensus regarding an appropriate definition of complexity (cf. Sect. 6.5). Some of the most convincing, albeit narrowly focused, evidence thereof has come from the painstaking study of the evolution of ammonoid sutures (Fig. 9.3; see also Boyajian and Lutz 1992), which rather convincingly reveals a gradual increase of complexity followed by degeneration (simplification) preceding extinction.

More recent and comprehensively quantitative work has explored the hypothesis that the accumulation of mildly deleterious mutations—which occurs according to Kimura's theory—leads to secondary selection for protein–protein interactions stabilizing key gene functions in small populations.[5] The argument of this work is that neutral drift of the genome and, in consequence, of the proteome leads to less stable proteins because of the occurrence of dehydrons (Sect. 11.5.2). The interactome (Chap. 16) is then developed to restore stability, which leads to the epiphenomenon of complexity. Lest it be thought that complexity is automatically a beneficial evolutionary trait, it should be pointed out that the prevalence of dehydrons in complex organisms such as ourselves leads to diseases due to unwanted protein aggregation such as Alzheimer's and is likely to increase the likelihood of aneuploidy and cancer. On the other hand this complexity appears to have been a prerequisite for the emergence of our brains with the concomitant ability to reflect on these matters and even, perhaps, find ways of overcoming the physiological drawbacks.

---

[3] E.g., Lynch (2007).
[4] McShea (1991).
[5] See Fernández and Lynch (2011) for full details.

Fig. 9.3 *Ammonites (Ptyichites) opulentus* (Mojsisovich), showing the complex sutures. Figure 5 from Plate 44 in E. Haeckel, *Kunstformen der Natur* (1. Sammlung). Leipzig: Verlag des Bibliographischen Instituts (1904). A collection of ammonoid fossils showing the increase and decrease of suture complexity can be viewed at the *Musée cantonal de zoologie*, Palais de Rumine, Lausanne

9.8   Self-Organization

The concept of self-organization appears to have originated with Immanuel Kant. In Sect. 65 of his *Kritik der Urteilskraft* (1790),[6] we read "In einem solchen Produkte der Natur wird ein jeder Teil, so wie er nur durch alle übrigen da ist, auch als um der anderen und das Ganzen willen existierend, d.i. als Werkzeug (Organ) gedacht; welches aber nicht genug ist (denn er könnte auch Werkzeug der Kunst sein und so nur als Zweck überhaupt möglich vorgestellt werden), sondern als ein die anderen Teile (folglich jeder den anderen wechselseitig) hervorbringendes Organ, dergleichen kein Werkzeug der Kunst, sondern nur der allen Stoff zu Werkzeugen (selbst denen der Kunst) liefernden Natur sein kann; und nur dann und darum wird ein solches Produkt als *organisiertes* und *sich selbst organisierendes Wesen ein Naturzweck* genannt werden können". The emphases, given by the author, seem to indicate his own feeling of the importance of this statement. As for the idea that a living organism is both cause and effect of itself, that is to be found in the preceding Sect. 64: "Ich würde vorläufig sagen: ein Ding existiert als Naturzweck, *wenn es von sich selbst Ursache und Wirkung ist* . . . ." Note Kant's caution in putting this forward as a provisional idea. Little, if anything, seems to have been added by the latterly often cited work of Maturana and Varela who introduced the term "autopoiesis"; they seem rather to have rendered a clear enough conception recondite. More constructive was the term "homeostasis" introduced by Cannon in the 1920s, and which became incorporated into Ashby's cybernetics (Sect. 9.4). Nevertheless, let us be mindful of Ashby's and von Foerster's critiques of self-organization (see footnote 25 in Sect. 2.4).

9.9   Cybernetics

In its modern incarnation, cybernetics was, initially, the study of control and communication within machines (considered as information processors, hence in this sense also encompassing living organisms), with the machine considered as an entity independent from the observer. Perhaps from the influence of quantum mechanics and its concept of the absolutely small quantum being irremeably perturbed by the observer, cyberneticians realized that they also needed to explicitly encompass the observer, and this extension of the earlier idea became known as "second-order cybernetics", or "cybernetics of cybernetics".

Chailakhian (2005) has pointed out the inevitability of "bioinformatics" becoming synonymous with "cybernetics" and it would be artificial to deny it. It is a *curiosum* that physiology still focuses on exchanges of energy rather than exchanges of information, although with the recognition of the importance of signalling within and between cells this is slowly changing. This synonymity vastly expands the scope of bioinformatics beyond genomics, especially when considering that control systems

---

[6]Quotations are taken from the third edition (Kant 1799).

are arranged hierarchically. "Survival" (Sect. 9.1) is a rather high level "goal". The action of breathing is quite central to it, but this implies not only operation of the autonomic and somatic nervous systems, but also whatever is needed to ensure that one is in a place where clean, respirable air is available, which itself implies myriads of actions, including some at the highest level of organization of society.

Problem.   Construct a concrete example of a hierarchical control system.

References

Boyajian G, Lutz T (1992) Evolution of biological complexity and its relation to taxonomic longevity in the Ammonoidea. Geology 20:983–986

Chaĭlakhian LM (2005) What is the subject of the science called bioinformatics? Biofizika 50:152–155 (in Russian)

Fernández A, Lynch M (2011) Non-adaptive origins of interactome complexity. Nature 474: 502–505

Kant I (1799) Critik der Urtheilskraft (3. Auflage). F.T. Lagarde, Berlin

Lipson H (2007) Principles of modularity, regularity, and hierarchy for scalable systems. J Biol Phys Chem 7:12–128

Lynch M (2007) The frailty of adaptive hypotheses for the origins of organismal complexity. Proc Natl Acad Sci 104:859–8604

McShea DW (1991) Complexity and evolution: what everybody knows. Biol Philos 6:30–324

Sommerhoff G (1950) Analytical biology. Oxford University Press, London

The Nature of Living Things

10

Figure 10.1 shows, in highly compressed and schematic form, the major processes taking place within living beings. The first priority of any living being is simply to survive. In the language of Sect. 9.4, the being must maintain its essential variables within the range corresponding to life. In succint form, "to be or not to be—that is the question."

The biosynthetic processes of life maintenance, indicated at the bottom of Fig. 10.1, lead beyond the living part of the organism to produce external structures, like exoskeletons and shells, which are sometimes gigantic, such as coral reefs, giant redwood tree trunks, guano hills, and, indeed, beaver dams and buildings of human construction.

Bioinformatics is particularly concerned with the processes of information flow (cf. the "central dogma"); that is, d, e, f in Fig. 10.1, and regulation of those processes (g, h, i). Nevertheless, any student of bioinformatics should have some grasp of the overall picture, which this chapter sets out to give.

The simplest organisms are single cells, slightly more elaborate organisms such as sponges consist of aggregates of cells constrained to live together, and more complex organisms are highly constrained assemblies of cells.

10.1 The Cell

The basic unit of life is the cell. Many organisms consist of only one cell. Therefore, even a single cell carries all that is needed for life. The cell contains the DNA coding for proteins and all the machinery necessary for maintaining life—enzymes, multiprotein complexes, and so forth. The body of the cell, the cytoplasm, is a thick, viscous aqueous medium full of macromolecules. If intact cells are centrifuged, one can separate a fairly fluid fraction, which contains very little apart from a few ions

© Springer-Verlag London 2015
J. Ramsden, *Bioinformatics*, Computational Biology 21,
DOI 10.1007/978-1-4471-6702-0_10

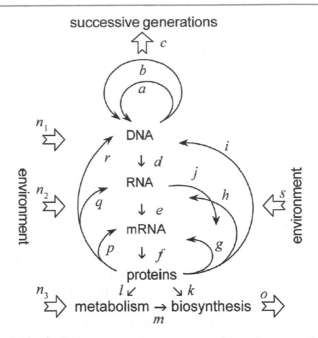

Fig. 10.1 Schematic diagram of the major relations in living organisms. The innermost zone concerns processes taking place within the cell. The *upper portion* indicates processes (a, b) involved in multiplication (reproduction); the *lower portion* indicates processes (k, l, m) involved in life maintenance (homeostasis). The curved arrows moving upward in the central region on the *left-hand side* indicate processes (p, q, r) of synthesis; those on the *right-hand side* (g, h, i, j) indicate processes of regulation. Exchange with the environment (input and output) takes place: Inputs n_1, n_2, and n_3 could be, respectively, cosmic rays causing DNA mutations, toxicants interfering with the regulation of transcription and translation, and food. s indicates specific molecular factors ingested from the environment, such as folic acid providing a source of methyl groups for DNA methylation. The successive generations (c) are, of course, released into the environment. Secretion (o) includes not only waste products but also highly specific molecules for altering a surface in the vicinity of an organism, or its outer shell (the set of secreted molecules other than waste is called the secretome)

and small osmolytes like sugars.[1] The proteins and the rest that are usually called "cytoplasmic" are mostly bound to macromolecular constructs such as the inner surface of the outer lipid membrane, internal membranes such as the endoplasmic reticulum, other polymers such as various filaments (the cytoskeleton) made from proteins such as actin or tubulin, or polysaccharides. These bound proteins can only be released if the ultrastructure of the cell is completely disrupted (e.g., by mechanically crushing it in a cylinder in which a tightly fitting piston moves); the results obtained from fractionating such homogenates give a quite misleading impression of the constitution of a living cell.

---

[1] Kempner and Miller (1968).

The cell membrane (also called "plasma membrane" or "plasmalemma"), often described as a robust and fairly impermeable coating around the cytoplasm, has a function that, strictly speaking, remains somewhat mysterious, since modern, and not so modern, research has shown that cells remain viable even when their membranes are significantly disrupted. The image of a cell as a toy balloon filled with salt solution, which would immediately spurt out if the balloon were punctured, is not in agreement with the experimental facts.[2]

10.1.1 The Structure of a Cell

The two great divisions of cell types are the prokaryotes (bacteria and archaeae) and the eukaryotes (protozoa, fungi, plants, and animals) (see Sect.10.9.5). As the name suggests, the eukaryotes possess a definite nucleus containing the genetic material (DNA), which is separated from the rest of the cell by a lipid-based membrane, whereas the prokaryotes do not have this internal compartmentation. Moreover, the eukaryotes possess other internal compartments known as organelles: the mitochondria, sites of oxidative reactions where food is metabolized; chloroplasts (only in plants), sites of photosynthesis; lysosomes, sacs of digestive enzymes for decomposing large molecules; the endoplasmic reticulum, a highly folded and convoluted lipid membrane structure to which the ribosomes (RNA-protein complexes responsible for protein synthesis from mRNA templates) are attached, and contiguous with the Golgi body, responsible for other membrane operations such as packaging proteins for excretion to outside the cell; and so on.

10.2 Mitochondria

The mitochondria and chloroplasts possess their own DNA, which codes for some, but not all of their proteins; they are believed to be vestiges of formerly symbiotic prokaryotes living within the larger eukaryotes. The present interrelationship between cell and mitochondrion is highly convoluted. The yeast mitochondrion, for example, has about 750 proteins, of which only 8 are templated by the mitochondrial genome, the remainder coming from the principal genome of the cell.

Mitochondria can undergo fusion and fission; dysfunction in these processes can lead to debilitating disease. Neurons seem to be particularly susceptible to mitochondrial dysfunction. Fusion is one way to rescue mitochondrial material that has lost function, and fission followed by elimination of the fragments is a way to eliminate irreparably damaged mitochondria.[3]

---

[2]See Kellermayer et al. (1986).
[3]Chan (2006) and Westermann (2010) are useful reviews.

10.2.1 Observational Overview

The optical microscope can resolve objects down to a few hundred nanometres in size.[4] This is sufficient for revealing the existence of individual cells (Hooke 1665) and some of the larger organelles (subcellular organs) present in eukaryotes. The contrast of most of this internal structure is low, however, and stains must be applied in order to clearly reveal them. Thus, the nucleus, chromosomes, mitochondria, chloroplasts, and so on can be discerned, even though their internal structure can not. The electron microscope, capable of resolving structures down to subnanometre resolution, has vastly increased our knowledge of the cell, although it must always be borne in mind that the price of achieving this resolution is that the cell has to be killed, sectioned, dehydrated or frozen, and stained or fixed—procedures that are likely to alter many of the structures from their living state.[5] Mainly through electron microscopy, a large number of intracellular structures, such as microfilaments, microtubules, endoplasmic reticulum, Golgi bodies, lysosomes, peroxysomes, and so on acquired something apparently more substantial than their previous somewhat shadowy existence.

If cells are mechanically homogenized, different fractions can be separated in the centrifuge: lipid membrane fragments, nucleic acids, proteins, polysaccharides, and a clear, mobile aqueous supernatant containing small ions and osmolytes. It should not be supposed that this supernatant is representative of the cytosol, the term applied to the medium surrounding the subcellular structures; centrifugation of intact cells (the experiments of Kempner and Miller 1968) removes practically all macromolecules along with the lipid-based structures. That experiment was done relatively late in the development of biochemistry, after the misconception that the cytosol was filled with soluble enzymes had already become established. Most proteins are attached to membranes, and the cytosol is a highly crowded, viscous hydrogel.[6]

Lipid membranes occupy a very important place in the cell. Most of the organelles are membrane-bounded, and their surfaces are the sites of most enzyme activity. Chloroplasts are virtually filled with internal membranes. Curiously, the most prominent membrane of all, the one which surrounds the cell, has, even today, a rather obscure function; for example, it is often asserted that it is needed to control ion fluxes

---

[4]According to Abbe's law, the resolution $\Delta x = \lambda/2(\text{N.A.})$, where λ is the wavelength of the illuminating light and N.A. is the numerical aperture of the microscope condenser. This barrier has now been broken by some remarkable new techniques developed by S.W. Hell, notably stimulated emission depletion (STED) and ground state depletion (GSD) microscopies, based on reversible saturated optical fluorescence transitions (RESOLFT) between two states of a fluorescent marker, typically a dye introduced into the living cell. The resolution is approximately given by $\Delta x_{\text{Abbe}}/\sqrt{1 + I/I_{\text{sat}}}$, where I is the actual illuminating irradiance and I_{sat} is the irradiance needed to saturate the transition.

[5]See Hillman (1991) for an extended discussion.

[6]See Ellis (2001).

into and out of the cell but, experimentally, potassium flux is much less affected by removal of the membrane than one might suppose (see footnote 2).[7]

Although prokaryotes (which are typically much smaller than eukaryotes) lack most of the internal membrane-based structure seen in eukaryotes, they are still highly heterogeneous in terms of the nonuniform distributions of components, from macromolecules down to small ions.

If they are tagged (by synthesizing them with unusual isotopes, or attaching a fluorescent label or a nanoparticle), individual molecules, or small groups of molecules, can be localized in the cell, by spatially resolved secondary ion mass spectrometry (SIMS), fluorescence microscopy, and so forth. These measurements can usually be carried out with fair time resolution (milliseconds to seconds); hence, both local concentrations and fluxes of the tagged molecules can be determined.

Spontaneous assembly. Take the isolated constituents (e.g., head, neck, legs) of a phage virus, mix them together, and a functional virus will result.[8] This exercise cannot be repeated successfully with larger, more complex structures closer to that state we call "living." Nor does it work if we break down the phage constituents into individual molecules.

10.3   Metabolism

The fundamental purpose of metabolism is to provide energy for survival (thinking, mobility, repair) and components for growth (including the production of offspring). It may be defined as the set of chemical reactions needed to maintain life—to grow, reproduce, repair, and respond (adapt). Traditionally, it is subdivided into catabolism, concerned with breaking large, usually polymeric, molecules imported as food from the external world into the cell down into monomers and submonomeric components in order to provide energy, and anabolism, concerned with building up large molecules and supramolecular structures. Metabolism is largely carried on by enzymes and coenzymes, the latter being molecules auxiliary to enzyme action that transfer chemical functional groups (e.g., $NAD^+/NADH$).

Digestion is typically carried out extracellularly (in the mouth, stomach and gut— the gastro-intestinal tract) and breaks macromolecular food (proteins, polysaccharides, fats) into oligomers that can be imported into the cell. The fundamental process of carbohydrate catabolism is glycolysis, which yields an intermediate molecule called pyruvate ($C_3H_4O_3$). Glycolysis is a principal energy source for prokaryotes and eukaryotes lacking mitochondria (e.g., erythrocytes). Within mitochondria, pyruvate is further broken down into acetyl coenzyme A (acetylcoA), which undergoes final decomposition in the citric acid (or tricarboxylic or Krebs) cycle, yielding two molecules of ATP (from ADP), one CO_2, and one NADH (from NAD^+). Oxygen

---

[7]Solomon (1960).
[8]See Kellenberger (1972) for a review.

is then used to regenerate the NAD^+ and a further molecule of ATP from ADP, together with a proton that is pumped outside the mitochondrion. The resulting proton electrochemical potential gradient across the mitochondrial membrane ("proton-motive force", p.m.f.) drives ATP synthase upon relaxation. This is called oxidative phosphorylation (respiration). It uses an exogenous electron acceptor (oxygen) to generate significant quantities of stored chemical potential ("energy"; more than 20 molecules of ATP per glucose molecule). Fermentation is an anaerobic process for further oxidizing pyruvate using an endogenous electron acceptor such as some other organic compound (lithotrophs use minerals), which yields much less stored chemical potential per glucose molecule than oxidative phosphorylation, perhaps only one-twentieth as much, depending on the final products. Photosynthetic organisms use light to reduce water to oxygen and develop a p.m.f. that is a similarly used to drive ATP synthesis across the thylakoid membrane.

Autotrophs such as plants can use the smallest carbon building block, namely CO_2, for anabolism, whereas heterotrophs use monomers for building up their catalytic and structural polymers.

Biological reactions, especially those *in vivo* within a cell, typically take place in very confined volumes. This confinement may have a profound effect on the kinetic mass action law (KMAL). Consider the reaction $A + B \xrightarrow{k_a} C$, which Rényi (1953) has analysed in detail. We have

$$\frac{dc}{dt} = k_a[\bar{a}\bar{b} + \Delta^2(\gamma_t)] = k_a\overline{ab} , \qquad (10.1)$$

where lower case symbols denote concentrations, bars denote expected numbers, and γ_t is the number of C molecules created up to time t. The term $\Delta^2(\gamma_t)$ expresses the fluctuations in γ_t: $\overline{\gamma_t^2} = \overline{\gamma_t}^2 + \Delta^2(\gamma_t)$. Supposing that γ_t approximates to a Poisson distribution, then $\Delta^2(\gamma_t)$ will be of the same order of magnitude as $\overline{\gamma_t}$. The KMAL, which puts $\bar{a} = a_0 - c(t)$, and so on, the subscript 0 denoting initial concentration (at $t = 0$), is a first approximation in which $\Delta^2(\gamma_t)$ is supposed negligibly small compared to $\bar{a}$ and $\bar{b}$, implying that $\bar{a}\bar{b} = \overline{ab}$, whereas, strictly speaking, it is not since a and b are not independent: the disappearance of A at a certain spot (i.e., its transformation into C) implies the simultaneous disappearance of B. The neglect of $\Delta^2(\gamma_t)$ is justified for molar quantities of starting reagents,[9] but not for reactions in minute subcellular compartments. The number fluctuations (i.e., the $\Delta^2(\gamma_t)$ term) will constantly tend to be eliminated by diffusion. This generally dominates in macroscopic systems. When diffusion is hindered, however, because of the correlation between a and b, initial inhomogeneities in their spatial densities lead to the development of zones enriched in either one or the other faster than the enrichment can be eliminated by diffusion.

Problem.   What fundamental limitations do small systems place on biological processes such as gene regulation?

---

[9]Except near the end of the process, when $\bar{a}$ and $\bar{b}$ become very small.

10.4   The Cell Cycle

Just as exponential decay is an archetypical feature of radioactivity, so is exponential growth an archetypical feature of the observable characteristics of life. If a single bacterium is placed in a rich nutrient medium, after a while (as little as 20 minutes in the case of *Escherichia coli*) two bacteria will be observed; after another 20 minutes, four, and so on; that is, the number n of bacteria increases with time t as e^t (cf. Eq. 7.4).

Actually, exponential growth, as known to occur under laboratory conditions, is not very common in nature. The vast majority of bacteria in soils and sediments live a quiet, almost moribund existence, due to the scarcity of nutrients. Under transiently favourable conditions, growth might start out exponentially but would then level off as nutrients became exhausted (cf. Eq. 7.5).

Bacteria "multiply by division." Since the average size of each individual bacterium remains roughly constant averaged over long intervals,[10] what actually happens is that the first bacterium increases in size and then divides into two. In general, the division does *not* appear to be symmetrical[11]—in other words, to express the result of the division as "two daughter cells" may not be accurate; there is a mother and daughter, and they are not equivalent.[12]

During the growth process, most of the molecules of the cell are increasing (in number) *pro rata* with overall cell size (mass), including the cell's gene, a circle of double-stranded DNA. Once the gene has been duplicated, the rest of the material can be divided, and growth starts again. The process has a cyclic nature and is called the cell cycle (Fig. 10.2).

The defining events are: the initiation of chromosome replication; chromosome segregation; cell division; and inactivation of the replication machinery. The duration of one cycle can vary by many orders of magnitude: 20 minutes for *E. coli* grown in the laboratory, to several years for the bacteria believed to live in deep ocean sediments. Typically, fully differentiated cells never divide.

The successive steps of the cell cycle appear to be tightly controlled and, if the control goes awry, damage and subsequent developmental abnormalities such as the formation of tumours may ensue. Control takes place principally at the checkpoints (corresponding to the boundaries separating the phases; Fig. 10.2) at which intervention is possible. Proteins called cyclins are synthesized just before each checkpoint is reached. They activate kinases that, in turn, phosphorylate other proteins

---

[10] E.g., Wakamoto et al. (2005).

[11] See Lechler and Fuchs (2005).

[12] The events of growth and division are not really akin to printing multiple copies of a book, or photocopying pages. It is not, strictly speaking, correct to call the process whereby adult organisms create new organisms—offspring—"reproduction": Parents do not reproduce themselves when they make a baby; even when the baby is grown up, it might be quite different, in appearance and behaviour, from its progenitors. In a literary analogy, this kind of process is akin to writing a new book (a derivative work) by gathering material from primary sources, or previously existing secondary sources.

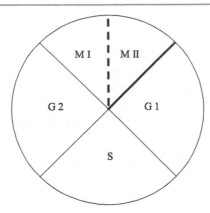

Fig. 10.2 Schematic diagram of the cell cycle. The successive epochs are known as phases. Areas of the sectors are proportional to the typical duration of each phase, which succeed each other in a *clockwise* direction. A newly born cell starts in the so-called G1 phase. When it reaches a certain size (the molecular nature of the initiating signal is not known, but it is correlated with size) DNA synthesis begins; that is, the gene is duplicated. Mitosis (see below) takes place in the M phase. See also Table 10.1

("cyclin-dependent kinases" (CDK), cf. Sect. 14.7) that carry out the necessary reactions to enable the cell to pass into the next phase, whereupon the cyclins are abruptly destroyed.

Apart from duplicating its DNA and dividing, the cell also has to metabolize food (to provide energy for its other activities, which may include secreting certain substances, or simply playing a structural rôle) and neutralize external threats such as viruses, toxins and changes in temperature. All of these activities, including gene duplication, require enzymes, and enzymes for translating and modifying the nucleic acid genetic material, whose fabrication also requires energy. There is also a considerable amount of degradation activity (i.e., proteolysis of enzymes and other proteins after they have carried out their specific function).[13] Degradation itself, of course, requires enzymes to carry it out. In eukaryotes, most proteins are marked for degradation by being covalently bound to one or more copies of the polypeptide ubiquitin. This facilitates their recognition by a huge ($M_r \sim 10^6$) multiprotein complex called the proteasome, which carries out proteolysis into peptides, which may be presented to the immune system, and ultimately to amino acids.

---

[13] A good example of a protein subjected to degradation is cyclin, which has the regulatory function mentioned above and whose concentration rises and then falls during mitosis.

10.4.1 The Chromosome

In eukaryotes, the nucleic acid is present as long linear segments (each containing thousands of genes) called chromosomes, because they can be coloured (stained) and hence rendered visible in the optical microscope during cell division.

Chromosomes are terminated by telomeres. The telomere is a stretch of highly repetitive DNA. Since during chromosome replication (see below) the DNA polymerase complex typically stops several hundred bases before the end, telomeres prevent the loss of possibly useful genetic information.

Germline cells are haploid; that is, they contain one set of genes (like bacteria). When male and female gametes (eukaryotic germline cells) fuse together, the zygote, the single-celled progenitor of the adult organism, therefore contains two sets of genes (i.e., two double helices), one from the male parent and one from the female parent. This state is called diploid. The normal descendents of the zygote, produced by mitosis, remain diploid. Many plants, and a few animals, have more than two sets of genes (four = tetraploid, many = polyploid), widening the possibilities for the regulation of gene expression. Polyploidy is a macromutation that greatly alters the biochemical, physiological, and developmental characteristics of organisms. It may confer advantageous tolerance to environmental exigency (especially important to plants (Sect. 20.2) because of their immobility) and open new developmental pathways. Cancers are characterized by aneuploidy, which, typically, leads to unpredictable further development. The unpredictability is autocatalytic, since once entire chromosomes are missing or duplicated, there is a chance at some of the machinery for copying the DNA is affected (cf. Sect. 10.6.3).

The two (or more) forms of the same gene are called alleles. The inheritance of unlinked genes (i.e., genes on different chromosomes; genetic linkage refers to the association of genes by virtue of their being located on the same chromosome) follows Mendel's laws.[14] If, for a given gene, two alleles are known, denoted A and a, occurring with probabilities p and $1 - p = q$, respectively, there are three possible genotypes in the population (AA, Aa, and aa), with probabilities of occurrence of p^2, $2pq$, and q^2, respectively (this is the Hardy–Weinberg rule). The Aa genotype is called heterozygous (the two parental alleles of a gene are different).

The union of a maternal and a paternal gene is typical of eukaryotes, a corollary of which is that siblings share half their genes with each other. The social insects are an important (recall that ants may comprise about a quarter of the animal mass on earth) exception. The queen is only fertilized once in her lifetime, storing the sperm in her body. She lays two kinds of eggs: fertilized with the stored sperm just before laying, and which become females; and unfertilized, which become males. The males therefore have only one set of chromosomes (i.e., they are haploid); in a

---

[14] 1. Phenotypical characters depend on genes. Each gene can vary, the ensemble of variants being known as alleles. In species reproducing sexually, each new individual receives one allele from the father and one from the mother. 2. When an individual reproduces, it transmits to each offspring the paternal allele with probability 1/2 and the maternal allele with probability 1/2. 3. The actual transmission events are independent for each independently conceived offspring.

Table 10.1 Successive events in the eukaryotic cell cycle

Phase[a]	Process	Feature(s)
M	Prophase	Chromosome condensation
M	Metaphase	Centrosomes separate and form two asteriated poles at opposite ends of the cell
M	Prometaphase	The nuclear envelope[b] is degraded, microtubules from the centrosomes seek the chromosomes
M	Metaphase	Microtubules from the centrosomes find the chromosomes
M	Anaphase A	The two arms of each chromosome are separated and drawn toward the centrosomes
M	Anaphase B	Centrosomes move further away from each other together with their half-chromosomes
M	Telophase	The cell divides
G1	Decondensation	Chromosomes disappear, nuclear envelope reforms around the DNA, microtubules reappear throughout the cytoplasm
S	Interphase	Cell growth
G2[c]	Interphase	DNA duplication

[a]See Fig. 10.2
[b]The nuclear envelope is a bilayer lipid membrane in which proteins are embedded
[c]Mitosis (see Sect. 10.4.1) is considered to begin at the end of G2 and last until the beginning of G1

certain sense, the males have no father. Hence, they transmit all their genes to their progeny, which are invariably female. In consequence, sisters share three-quarters of their genes with each other, but they only have a quarter of their genes in common with their brothers.[15]

Mitosis

The simple process of gene replication is called mitosis. This is the type of cell division that produces two genetically identical (in theory) cells from a single parent cell. It applies to bacteria and to the somatic (body) cells of eukaryotes.

Prior to division, homologous pairs (of the maternal and corresponding paternal genes for each chromosome) form. They are attached at one zone, near the centre of the chromosome, by a large multiprotein complex called the centromere. The attached chromosomes then compactify, forming the characteristic "X"-shaped structures easily seen in the optical microscope after staining. The remainder of the process is described in Table 10.1.

---

[15]This fact is used to "explain" social insect behaviour.

Meiosis

Meiosis is a more complex process than mitosis. It starts with an ordinary diploid cell and leads to the formation of gametes (germline cells).

First, the two chromosomes (paternal and maternal) are duplicated (as in mitosis) to produce four double helices. Then the four double helices come into close proximity and recombination (see below) is possible. Thereupon the cell divides without further DNA replication. The chromosomes are segregated; hence, each cell contains two double helices (diploid). A given double helix may have sections from the father and from the mother. Finally, there is a further division without further DNA replication. Each cell contains one double helix (haploid). They are the gametes (germ cells).

Differences Between Prokaryotes and Eukaryotes (1)

Prokaryotes undergo neither meiosis nor mitosis (their DNA is segregated as it replicates), their chromosomes are not organized into chromatin (although there is a region called the nucleoid in which the genetic material is concentrated), nor does the DNA spend much of its time inside a special compartment, the nucleus (although the chromosome is usually visible as the nucleoid). Chromosome replication typically starts from a single site in prokaryotes (the origin of replication, *ori*, which may comprise a few hundred bases) but from many sites (thousands) in eukaryotes—otherwise replication, proceeding at about 50 bases per second, would take far too long. As it is, the human genome takes about 8 h to be replicated. Prokaryotic DNA is circular (and hence does not require telomeres),[16] whereas eukaryotic DNA is linear.

Differences Between Protozoans and Metazoans

In a single-celled protozoan, the germline is the soma (body). The metazoan is quite different because its germline (a single cell) must divide and multiply in order to create the soma. All cells have the same genes (with some specialized exceptions, such as in the cells of the immune system; cf. Sect. 10.5). Typically, methylation of the DNA determines which genes are expressed; in the germ cell, only "master control genes" are unmethylated; these control the demethylases, which progressively allow other genes to be expressed. As a rule, this development takes place under much more strongly constrained environmental conditions than those that the fully developed (adult) organism might expect to encounter. Imprinted genes are those whose expression is determined by their parental origin, typically according to the genes' methylation states (see Sect. 10.7.4).

---

[16]There are some exceptions; for example, *Streptomyces coelicolor* has a linear genome.

10.4.2 The Structures of Genome and Genes

Definition. We may provisionally define *gene* as a stretch of DNA that codes for (i.e., is translated into—see Sect. 10.7) a protein. Due to ever more detailed molecular knowledge, it has become difficult to define "gene" unambiguously. Formerly, the term "cistron" was used to denote the genetic unit of function corresponding to one polypeptide chain; the discovery of introns (see below) signified the end of the "one gene, one enzyme" idea; furthermore, operons group together several proteins with a common function—are they then to be regarded as a single gene? The genon concept (see below) may provide a way of reconciling the classical view of a gene as a function and the molecular biological view of the gene as a coding sequence (with the ambiguity of whether to include sequences involved in regulating expression).

Definition. The *genon* has been introduced by Scherrer and Jost (2007) in an attempt to delineate an object that can be defined unambiguously. The genon is defined as the coding sequence (which can then revert to being called "gene," akin to the sense of cistron, but better (less ambiguously) expressed in terms of the mRNA that is translated into a protein) together with the additional information that is needed to fully express the coding sequence. The genon is therefore more akin to a program that results in a functionally active gene product. The coding sequence together with its promoter is called the protogenon, and the primary transcript is called the pregenon. These are comprised within the cisgenon, together with RNA and proteins necessary for expression. Once the protein is produced, we move into the domain of the transgenon, which finally denotes the working protein delivered at a particular time to a particular place in the cytoplasm. Doubtlessly, this concept will be further refined and its operational implications more fully explored.

Definition. The *genome* is defined as the entire set of genes in the cell. Intergenomic sequences and introns (a term suggested by Walter Gilbert in 1978, signifying intragenic sequences) were not known when the word was coined. Therefore it is usually taken to mean all inheritable polymerized nucleic acids, regardless of their coding or other function.

The most basic genome parameter is the the number of bases (base pairs, since most genetic DNA is double stranded). Sometimes the molecular weight of the DNA is given (the average molecular weight of the four base pairs is 660). Table 10.2 gives the sizes of the genomes of some representative organisms.

Differences Between Prokaryotes and Eukaryotes (2)

Bacterial genomes consist of blocks of genes preceded by regulatory (promoter) sequences. Eukaryotic DNA resembles a mosaic of the following: genes (segments whose sequence codes for amino acids, also called exons, from *ex*pressed, or "coding DNA"); segments (called introns) that are transcribed into RNA, but then excised to leave the final mRNA used as the template for producing the protein; many genes are split into a dozen or more segments, which can be spliced in different ways to generate variant proteins after translation; promoters (short regions of DNA to which RNA, proteins, or small molecules may bind, modulating the attachment of RNA

Table 10.2 Some genome data

Organism	Number of base pairs (bp) per haploid cell	Number of genes	Haploid number of chromosomes	Number of cell types (approx.)
Escherichia coli	4×10^6	4290	1	1
Streptomyces coelicolor	8.6×10^6	7830	1	2
Amoeba dubia	7×10^{11}	60000	~300?	1
S. cerevisiae	10^7	6300	16	2
C. elegans	9×10^7	19000	6	30
D. melanogaster	1.8×10^8	13500	4	50
Oikopleura dioica[a]	7×10^7	15000	8	?
Protopterus (lungfish)	1.4×10^{11}	?	~17	?
Triturus (newt)	1.9×10^{10}	?	12	150
O. anatinus[b]	3.06×10^9	18500	26	?
Mus musculus	3.5×10^9	30000	20	90
Homo sapiens	3.5×10^9	30000	23	220
Neurospora crassa	4×10^7	10000	7	28
D. discoideum[c]	3.4×10^7	12500	6	4
Ophioglossum (fern)	?	?	200–500	?
Arabidopsis thaliana	1.35×10^8	25500	5	?
Fritillaria[d]	1.3×10^{11}	?	12	?
Picea abies (Norway spruce)	2×10^{10}	28350	12	?

[a] A tunicate
[b] The duck-billed platypus
[c] Slime mould
[d] A bulbous plant from the Liliaceae family, not to be confused with the small mesopelagic larvaceans in the genus *Fritillaria*, nor with the fritillary, the name given to several species of butterfly from the subfamily Heliconiinae

polymerase to the start of a gene); and intergenomic sequences (the rest, sometimes called "junk" DNA in the same sense in which untranslated cuneiform tablets may be called junk—we do not know what they mean). This is schematically illustrated in Fig. 10.3.

Although the DNA-to-protein processing apparatus involves much complicated molecular machinery, some RNA sequences can splice themselves. This autosplicing capability enables exon shuffling to take place, suggesting the combinatorial assembly of exons *qua* irreducible codewords as the basis of primitive, evolving life.

Organisms other than prokaryotes vary enormously in the proportion of their genome that is not genes. The intergenomic material may exceed by more than an order of magnitude the quantity of coding DNA. Some of the intergenomic material is specially named, notably repetitive DNA. The main classes are the short (a few hundred nucleotides) interspersed elements (SINES), the long (a few thousand

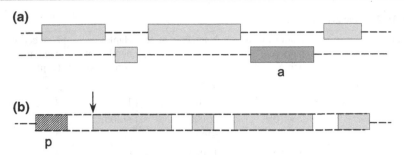

Fig. 10.3 Simplified schematic diagram of eukaryotic gene structure. **a** is the antiparallel double helix. *Rectangles* represent genes and *dashed lines* represent intergenomic sequences. **b** is an expansion of *a* (a single gene) in **a**. The *shaded rectangles* correspond to DNA segments transcribed into RNA, spliced, and translated continuously into proteins. *p* is a promoter sequence. In reality, this is usually more complex than a single nucleotide segment; it may comprise a sequence to which an activator protein can bind (the promoter site proper) but, also, more distant ("upstream" from the gene itself), one or more enhancer sites to which additional transcription factors (TF) may bind. All of these segments together are called the transcription factor binding site (TFBS). There may be some DNA of indeterminate purpose between *p* and the transcription start site (TSS) marked with an *arrow*. Either several individual proteins bind to the various receptor sites, and are only effective all together, or the proteins preassociate and bind *en bloc* to the TFBS. In both cases, one anticipates that the conformational flexibility of the DNA is of great importance in determining the affinity of the binding. To the *right* of the TSS: *shaded regions*, exons; *unshaded regions*, introns

nucleotides) interspersed elements (LINES), and the tandem (i.e., contiguous) repeats (minisatellites and microsatellites,[17] variable-length tandem repeats (VNTR) etc.).[18] These features can be highly specific for individual organisms. Several diseases are associated with abnormalities in the pattern of repeats; for example, patients suffering from X syndrome have hundreds or thousands of repeated CGG triplets at a locus (i.e., place on the genome) where healthy individuals have about 30. The rôle of repetition in DNA is still rather mysterious. One can amuse oneself by creating sentences such as "can a perch perch?" or "will the wind wind round the tower?" or "this exon's exon was mistranslated"[19] to show that repetition is not necessarily nonsense. The genome of the fruit fly *Drosophila virilis* has millions of repeats of three satellites, ACAAACT, ATAAACT and ACAAATT (reading from the 5′ to the 3′ end), amounting to about 10^8 base pairs (i.e., comparable in length to the entire genome, which does not exceed 2×10^8 base pairs). Another kind of repetition occurs as

---

[17] So called because their abnormal base composition, usually greatly enriched in C-G pairs (CpG), results in satellite bands appearing near the main DNA bands when DNA is separated on a CsCl density gradient.

[18] Archaeal and bacterial genomes contain clustered regularly interspaced short palindromic repeats (CRISPR; see, e.g., Sander and Joung 2014). They have found technological application as a way of genome editing.

[19] Most English dictionaries give only one meaning for exon, namely one of four officers acting as commanders of the Yeomen of the Guard of the Tower of London.

the duplication in the sense of further multiplication of whole genes along the chromosome (or on another chromosome). The apparently superfluous copies tend to acquire mutations vitiating their ability to be translated into a functional protein, whereupon they are called pseudogenes.[20] Gene duplication may be considered as a form of cellular computing.[21] In the human, satellite sequences of repetitive DNA alone constitute about 5 % of the genome; in the horse, they constitute about 45 %. Telomere sequences are further examples of repetitive DNA (in humans, TTAGGG is repeated for 3–20 kb). Between the telomere and the remainder of the chromosome there are 100–300 kb of telomere-associated repeats.

10.4.3 The C-Value Paradox

Well before genome sequence information became available, it was clear that the amount of DNA in an organism's cells (the C-value; more precisely it is the mass of DNA within a haploid nucleus) did not correlate particularly well with the organism's complexity, and this became known as the "C-value paradox." Examination of Table 10.2 will reveal some striking instances—the genomes of amoebae and lungfish considerably exceeding in size those of ourselves, for example.

Before delving into this question more deeply, three relatively trivial factors affecting the C-value should be pointed out. The first is experimental uncertainty, and ambiguity in the precise definition of the C-value. Second, in some cases, genome size is merely estimated from the total mass of DNA in a cell. This makes the given value highly dependent on polyploidy, unusual in mammals but not in amphibians and fish, and rather common in plants. For example, the lungfish, which has a conspicuously large C-value, is known to be tetraploid. Amoebae, which apparently have an even larger C-value, are likely to be polyploid and, moreover, the amount of DNA found in an amoeba cell may well be inflated by the remains of genetic material of recently ingested prey. Care should therefore be taken to ascertain the amount of genetic material corresponding to the haploid genome for the purposes of comparison. The third factor is the presence of enormous quantities of repetitive DNA in many eukaryotic genomes. These repetitive sequences include retrotransposons, vestiges of retroviruses, and so forth. Probably about half of the human genome can be accounted for in this way, and it seems not unreasonable to consider this as "junk" (although it appears to play a rôle in the condensation of the DNA into heterochromatin; see Sect. 10.4.4).[22]

---

[20] For a concrete example see Hittinger and Carroll (2007).

[21] Shapiro (1992).

[22] Regarding the remainder, about 5 % is considered to be conserved (by comparison with the mouse); 1.2 % is estimated to be used for coding proteins, and the remaining 3.8 % is referred to as "noncoding," although conservation of sequence is taken to imply significant function (it seems very probable that this "noncoding" DNA is used to encode the small interfering RNA used to

Is There a G-Value Paradox?

By correcting for polyploidy and repetitive junk, one arrives at the quantities of
DNA involved in protein synthesis (both the genes themselves and the regulatory
overhead). In some cases, the actual number of genes (the G-value) can be estimated
with reasonable confidence; in other cases, simple application of a compression
algorithm (Sect. 3.4) can be used to provide a minimal description (an approximation
to the algorithmic information content; see Chap. 6), which correlates much better
with presumed organismal complexity (as measured, for example, by the number of
different cell types). Where gene number estimates are available, however, the more
complex organisms do not seem to have enough genes. Especially if the figure for *H.
sapiens* has to be revised downward to a mere 20 000, we end up with fewer genes
than *A. thaliana*, for example! This is the so-called G-value paradox. Its resolution
would appear to lie with enhanced alternative splicing possibilities for more complex
organisms. We humans appear to have the largest intron sizes, for example.[23]

Differences Between Prokaryotes and Eukaryotes (3)

The above considerations do not directly address the question of why prokaryotes
have rather compact genomes; they seem to be limited to about 10 million base
pairs (10 Mb) (and many bacteria living practically as symbionts in a highly constrained environment manage with far less). In a general sense, one can understand
that prokaryotes are under pressure to keep their genomes as small as practicable;
they are usually replicating rapidly under r-selection (Sect. 10.8.4) and the need to
copy 1000 million base pairs would be physicochemically incompatible with a short
interval from generation to generation. On the other hand, most of the cells in a
metazoan are not replicating at all, and the burden of copying enormous genomes
during development is perhaps compensated for by the availability of plenty of raw
material for exploratory intraorganismal gene development (which the prokaryotes
do not need because of the facility with which they can acquire new genetic material
from congeners).

It has recently been shown that the nature of gene regulation also imposes certain
constraints on the relationship between the amounts of DNA assigned to coding
(for proteins) and those which are considered to be noncoding (i.e., corresponding to
regulatory sites such as promoters). According to what is known about the molecular
details of gene transcription (Sect. 10.7.2), to a first approximation each gene (with
an average length of about 300 base pairs) requires a promoter site (which might
have of the order of 10 base pairs). This gives 9:1 as the typical ratio of "coding" to

---

(Footnote 22 continued)
supplement protein-based transcription factors as regulatory elements). That still leaves the enigma
of the remaining 40–50 % that is neither repetitive nor coding in any sense understood at present.
[23]Taft et al. (1992). Note the connexion between alternative splicing and Tonegawa's mechanism
for generating B-cell lymphocyte (and hence antibody) diversity in the immune system (Sect. 10.5).

"noncoding" DNA in prokaryotes.[24] In the spirit of Wright's "many to many" model of regulation, gene regulatory networks are expected to be of the "accelerated growth" type (see Sect. 7.2), because each new gene that is added should be regulatorily connected to a fixed fraction $\mathfrak{r}$ of the existing genes. Hence, if g is the number of genes, then the number of regulations (edges of the graph) $r = \mathfrak{r}g^2$. These regulations are themselves mediated by proteins (the transcription factors) encoded by genes. However, there is an upper limit to the number of interactions in which a protein can participate, roughly fixed by the number of possible binding sites on a protein and their variety; empirical studies[25] suggest that the upper limit k_{max} of the degree k of the network is about 14. Since $k = 2r/g$, this suggests $g_{max} = k_{max}/(2\mathfrak{r})$, which would appear to correspond to the 10^7 base pairs maximum genome size of prokaryotes.

As is well known, however, even allowing for possible overstatement in eukaryotic genome length, far larger eukaryotic genomes are known to occur. Given their evident regulatory success (as evinced by the real increase in organismal complexity), one may suppose that the "accelerated growth" network model still holds; that is, all of the additional proteins are properly regulatorily integrated. Ahnert et al. (2008) have proposed that the regulatory deficit implied by $g > g_{max}$ is met by "noncoding" RNA-based regulation (see Sect. 10.7.4), the overhead of which is much smaller than that of the protein (transcription factor)-based regulation; this is borne out by the length of "noncoding" DNA ($\propto r$) increasing quadratically with the length of coding DNA ($\propto g$) above the 10 Mb threshold. It begs the question of why protein-based regulation is used at all, even in prokaryotes, if the RNA-based system is effective and much less costly, but our present knowledge of RNA-based regulation seems to be too incomplete to allow this question to be satisfactorily addressed.

DNA Base Composition Heterogeneity

The base composition of DNA is very heterogeneous,[26] which makes stochastic modelling of the sequence (e.g., as a Markov chain) very problematical. This patch-iness or blockiness is presumed to arise from the processes taking place when DNA is replicated in mitosis and meiosis (Sect. 10.4.1). It has turned out to be very useful for characterizing variations between individual human genomes. Much of the human genome is constituted from "haplotype blocks," regions of about 10^4–10^5 nucleotides in which a few (< 10; the average number is 5.5) sequence variants account for nearly all of the variation in the world human population. The haplotype "map" is simply a list of the variants for each block (cf. Sect. 20.6).

Haplotypes are essentially long stretches of DNA characterized by a small number of single-nucleotide polymorphisms (SNPs—pronounced "snips"); that is, mutated nucleotides. There is an average of about 1 SNP per thousand base pairs in the human

---

[24]Some groups of genes, typically those related functionally (such as successive enzymes in a metabolic pathway), are organized into "operons" controlled by a single promoter site and are therefore transcribed together (see also Chap. 16).

[25]Kim et al. (2006).

[26]E.g., Karlin and Brendel (1993).

genome; thus, if they were uncorrelated, in a typical 50 000 base pair haplotype block there would be about 2^{50} (or 4^{50}, depending on whether we are interested in what the base is mutated to) variants—far more variation than is actually found. Hence, the pattern of SNPs evinces extremely strong constraint; that is, the occurrences of individual SNPs are strongly correlated with each other. There is considerable current interest in trying to correlate haplotype variants with disease, or propensity to disease.[27] This is discussed in more detail in Sect. 20.6.

One notes that as much as 98 % of the human genome may be identical with that of the ape; one could equally well state that there is more genetic difference between man and woman than between man and ape. To actually derive the vast phenotypic differences between the two from their genomes appears to be as vain a hope as solving the Schrödinger equation for even a single gene.

As an information-bearing symbolic sequence, the genome is unusual in that it can operate on itself. The most striking example is furnished by retrotransposons (i.e., transposable elements, whose existence was first proposed by McClintock 1950). These gene segments *inter alia* encode a reverse transcriptase enzyme, which facilitates the making of a DNA copy of the sequence. The duplicate sequence is then inserted into the genome; the point of insertion may be remote from that of the gene from which the copy was made. The basis for McClintock's proposal was her observation of rapid variation of the colours of maize kernels from one generation to another; the interpretation of these changes was that a gene coding for colour could be inactivated if a transposon were inserted within it, but the transposon could, with equal facility, be removed during the next round of meiosis, resulting in the reappearance of the colour.

10.4.4 The Structure of the Chromosome

DNA is subject to oxidation, hydrolysis, alkylation, strand breaks and so forth, countered by various repair mechanisms as discussed in Sect. 10.6.2. Molecular machinery (called "SOS") is available to allow replication to proceed despite lesions. Mistakes are the origin of the genotypic mutations leading to the phenotypic variety required by Darwin's theory (see Sect. 10.9).

Eukaryotic DNA is organized into chromatin, a protein-DNA complex. The fundamental unit of chromatin structure is the nucleosome, a spheroidal complex about 9 nm in diameter made up from eight proteins called histones, around which a stretch of 140–200 DNA base pairs is wrapped (recall that the DNA double helix is about 2 nm in diameter). The chromosome is constituted from successive nucleosomes, joined by short stretches of so-called linker (non-nucleosomal) DNA. The string of nucleosomes and their linkers are then compacted into fibres about 30 nm in diameter, and these in turn are compactly folded to form the so-called chromatin loops, about 300 nm in diameter, of the chromosome. Much of the DNA, perhaps as much as 90 % in a resting cell, is in this highly condensed, somewhat

---

[27] Another curiosity is that certain DNA sequences display extraordinarily long-range (10^4 base pairs or more) correlations (see, e.g., Voss 1992).

inert state called heterochromatin. The condensation appears to occur in association with long sequences of repetitive DNA. Hypermethylation of cytosine is typical (see Sect. 10.7.4). The active portion, available for transcription, is known as euchromatin.

The protein core of the nucleosome plays a highly significant rôle in the regulation of transcription (Sect. 10.7.2). The amino acids of the histones are subject to many modifications, such as a acetylation, methylation, phosphorylation, and ubiquitination. Hypoacetylation of lysine is associated with heterochromatin formation.[28] Methylation of specific lysines is also associated with heterochromatin (and silencing of euchromatin genes). It is important to bear in mind that histone modification is a highly dynamic process, constantly under adjustment. Furthermore, there is evidence that the histones are precisely positioned relative to the DNA according to its sequence.[29]

10.5 The Immune System

The higher metazoans have developed a sophisticated mechanism for neutralizing external attack at the micrometre and nanometre scales, at which the dangers are bacteria and other microbes, viruses and dust particles. This mechanism is called the immune system and is divided into innate and adaptive parts. The primary (initial) response is innate and consists in the ingestion of foreign microbodies and nanobodies by phagocytes. The adaptive immune system, which is only found in the highest organisms, involves T-cells (matured in the *t*hymus) that have thousands of copies of the so-called T-cell receptor (TCR) on their surfaces. In principle, each T-cell has a different TCR, and each one can bind to (i.e., is a receptor for) a particular oligopeptide–MHC complex,[30] provided that the peptide is not from one of the organism's proteins. As a result of the binding, these "helper" T-cells release cytokines, which themselves trigger the proliferation and recruitment of cytotoxic or killer T-cells, which release perforin, a protein that perforates the target cell membrane when it binds to it. At the same time, B-cells (matured in the *b*one marrow) produce antibodies able to bind to portions ("antigens") of the foreign objects. Each B-cell produces a unique antibody.[31] The binding of a B-cell to "its" antigen leads to clonal expansion of that B-cell and concomitant expansion of antibody production. The antibodies binding to the antigens of the foreign objects form molecular complexes that are also recognized by T-cells, a process that leads to the destruction of

---

[28] See, for example, Jenuwein and Allis (2001), and Richards and Elgin (2002).

[29] Audit et al. (2002).

[30] The MHC (major histocompatibility complex) is a system of proteins residing on the surface of a cell that complexes with certain oligopeptides derived from a sample of the internal proteins of the cell.

[31] The antibody is a protein made up from several different polypeptide chains. Part of the molecule is the same for all antibodies and part is unique. Tonegawa (1983) demonstrated that the diversity of antibodies was due to somatic generation of genetic diversity among the genes coding for the variable (unique) part.

(or the attempt to destroy) the foreign objects (since they "present" antigens to the T-cells, they are known as antigen-presenting cells (APC)).

The number N of foreign antigens that must be recognized by an organism is very large, perhaps greater than 10^{16}, and at the same time there is a smaller number, $N' \sim 10^6$, of self-antigens that must *not* be recognized. Yet, according to Tonegawa's theory (see footnote 31), the immunoglobulin and T-cell receptors may only contain $n \sim 10^7$ different motifs. Recognition is presumed to be accomplished by a generalized lock-and-key mechanism involving complementary amino acid sequences. How large should the complementary region be, supposing that the system has evolved to optimize the task? If P_S is the probability that a random receptor recognizes a random antigen, the value of its complement $P_F = 1 - P_S$ maximizing the product of the probabilities that each antigen is recognized by at least one receptor and that none of the self-antigens is recognized (i.e., $(1 - P_F^n)^N P_F^{nN'}$) is[32]

$$P_F = \left(1 + \frac{N}{N'}\right)^{-1/n} . \tag{10.2}$$

Using the estimated values for n, N, and N', one computes $P_S \approx 2 \times 10^{-6}$. Suppose that the complementary sequence is composed of m classes of amino acids and that at least c complementary pairs on a sequence of s amino acids are required for recognition. Since the probability of a long match is very small, to a good approximation the individual contributions to the match can be regarded as being independent. A pair is thus matched with probability $1/m$ and mismatched with probability $1 - 1/m$. Starting at one end of the sequence, runs of c matches occur as with probability m^{-c}, and elsewhere they are preceded by a mismatch and can start at $s - c$ possible sites. Hence

$$P_S = [(s - c)(m - 1)/m + 1]/m^c . \tag{10.3}$$

If $s \gg c > 1$, one obtains

$$c = \log_m [s(m - 1)/m] - \log_m P_S . \tag{10.4}$$

Problem. Estimate c, Supposing s to be a few tens, $m = 3$ (positive, negative, and neutral residues), and using the numbers given above (since they all enter as logarithms the exact values are not critical). How does it compare with observation?

10.6   Molecular Mechanisms

In this section, DNA replication and recombination will be examined from the molecular viewpoint. The reader may find it useful to refer to Chap. 11 for complementary information.

---

[32]Percus et al. (1993).

Table 10.3 DNA replication

Name	Operand	Operation	Operator	Result
Premelting	Double helix	Facilitation	Topoisomerase	Strand separation
Melting	Double helix	Facilitation	Helicase	Strand separation
Synthesis	Single strand	Nucleotide addition	Polymerase	Semiconservatively replicated double helix

Two DNA polymerases are simultaneously active. They catalyse template directed growth in the $5' \rightarrow 3'$ direction. The leading strand is synthesized continuously from $5' \rightarrow 3'$ using the strand beginning with the $3'$ end as the template, whereas the lagging strand is synthesized in short ("Okazaki") fragments using the strand beginning with the $5'$ end as the template. A DNA primase produces a very short RNA primer at the $5'$ end of each Okazaki fragment onto which the polymerase adds nucleotides. The RNA is then removed by an RNAaseH enzyme. A DNA ligase links the Okazaki fragments. A set of initiator proteins is also required to begin replication at the origin of replication. This is, of course, a simplification; for example, it is estimated that almost 100 (out of a total of approximately 6000) genes in yeast are used for DNA replication, and another 50 are used for recombination

Table 10.4 Some types of chromosome rearrangements (with examples)

Name	Before[a]	After[a]
Deletion	ABCDEFGH	ABEFGH
Insertion	ABCDEFGH	ABCJFKDEFGH
Inversion	ABCDEFGH	ABCFEDGH
Transposition	ABCDEFGH	ADEFBCGH
Tandem duplications	ABCDEFGH	ABCBCBCDEFGGGGGH

[a]Each letter represents a block of one or more base pairs

10.6.1 Replication

The molecular mechanism of DNA replication is summarized in Table 10.3. Some of the typical errors—leading to single point mutations—that can occur are summarized in Table 10.4.

10.6.2 Proofreading and Repair

Many proteins are involved in the repair of mismatched, and breaks in, DNA. Repair takes place after replication, but before transcription. As with Hamming's error-correcting codes (Sect. 3.6), the DNA repair proteins must first recognize the error and then repair it. It is of primordial importance that DNA is organized into a double helix; the antiparallel strand can be used to check and template repair of mistakes recognized in the other one. Instead of repair, apoptosis (death of a single cell; as opposed to necrosis, death of many cells in a tissue) of the affected cell may occur. Concomitant with the work of the specific error recognition and repair enzymes, the entire cell cycle may need to be slowed to ensure that there is time for the repair work to be carried out. The mending systems are also used to repair damage caused by external factors (e.g., cosmic ray impact, oxidative stress etc.).

The available mechanisms are essentially directed toward repairing single-site errors; there is no special apparatus for eliminating gene duplications and the like. On the other hand, it is not only base mismatches that need to be repaired. Alkylation (methylation) damage could adversely affect gene expression and there are enzyme systems (oxidative demethylases and others) for repairing it.

Just as certain sequences are more prone to error than others, so are certain erroneous sequences more easily repaired than others. Whereas the quality of a telephone line is, essentially, independent of the actual words being said, the fidelity of DNA replication may be sequence-dependent. This possibility could be used by the genome to explore (via mutations) neighbouring (in sequence space) genomes. Hence, bioinformatics (applied to genomics) needs a higher-level theory than that provided by existing information theory. An important, although long-term, task of bioinformatics is to determine how biological genomes are chosen such that they are suited to their tasks, encompassing such aspects.

Unreliable DNA polymerase is a distinct advantage for producing new antibodies (somatic hypermutation) and for viruses needing to mutate rapidly in order to evade host defences—provided it is not too unreliable: Eigen (1976) has shown that in a soup of self-replicating molecules, there is a replication error rate threshold above which an initially diverse population of molecules cannot converge onto a stable, optimally replicating one (a quasispecies[33]).

Problem. What are the implications of a transcription error rate estimated as 1 in 10^5? (in contrast, the error rate of DNA replication is estimated as 1 in 10^{10}). Calculate the proportion of proteins containing the wrong amino acids due to mistakes in transcription, assuming that translation is perfect. Compare the result with a translation error rate estimated as 1 in 3000.

Problem. Explore the suggestion that the quality of a channel (such as a telephone line) is independent of the actual message.

10.6.3 Recombination

Homologous recombination is a key process in genetics, whereby the rearrangement of genes can take place. It involves the exchange of genetic material between two sets of parental DNA during meiosis (Sect. 10.4.1). The mechanism of recognition and alignment of homologous (i.e., with identical, or almost identical, nucleotide

---

[33] A quasispecies may be defined as a cluster of genomes in sequence space, the diameter of the cluster being sufficiently small such that almost every sequence can "mate" with every other one and produce viable offspring. The sequence at the centre of the cluster is called the master sequence. If the error rate is above the threshold, in principle all possible sequences will be found. See also Sect. 10.9.2.

sequences) sections of duplex (double-stranded) DNA is far less clear than the recognition between complementary single strands, but may depend on the pattern of electrostatically charged (ionized) phosphates, which itself depends slightly but probably sufficiently on sequence, and can be further modulated by (poly)cations adsorbed on the surface of the duplex.[34]

Following alignment, breakage of the DNA takes place, and the broken ends are then shuffled to produce new combinations of genes; for example, consider a hypothetical replicated pair of chromosomes, with the dominant gene written in majuscule and the recessive allele written in miniscule. If $*$ represents a chromosome break, we have for the duplex (concerning DNA strands numbered i to iv):

$$
\begin{array}{l|l}
i & ABC \\
ii & ABC \\
iii & abc \\
iv & abc
\end{array}
\rightarrow
\begin{array}{l|l}
i & AB * C \\
ii & A * BC \\
iii & a * bc \\
iv & ab * c
\end{array}
\rightarrow
\begin{array}{l|l}
i & ABc \\
ii & Abc \\
iii & aBC \\
iv & abC
\end{array}
. \qquad (10.5)
$$

There is supposed to be about one crossover per chromosome per meiosis. In more detail, the stages of recombination are the following:

1. Alignment of two homologous double-stranded molecules;
2. Breakage of the strands to be exchanged;
3. Approach of the broken ends to their new partners and formation of a fork (also known as a Holliday junction);
4. Joining of broken ends to their new partners;
5. Prolongation of the exchange via displacement of the fork;
6. End of displacement;
7. Breakage of the 3' extremities;
8. Separation of the two recombinant double strands;
9. Repair of the breaks via reading from the complementary strand.

The process is drawn in Fig. 10.4.

Unlike replication, in which occasional single-site ("point") mutations occur due to isolated errors, recombination results in changes in large blocks of nucleotides. Correlations between mutations greatly depend on the number of chromosomes. In species with few chromosomes, reshuffling is combinatorially limited and mutations in different genes are likely to be transmitted together from one generation to another, whereas in species with large numbers of chromosomes, randomization is more effective. There are also mechanisms whereby chromosome fission and fusion can occur, leading to aneuploidy (cf. Sect. 10.4.1), which is a hallmark of cancer (Sect. 20.2).

---

[34] Kornyshev and Leikin (2001).

Fig. 10.4 Strand exchange in homologous recombination. The *numbers* refer to the stages described in the text

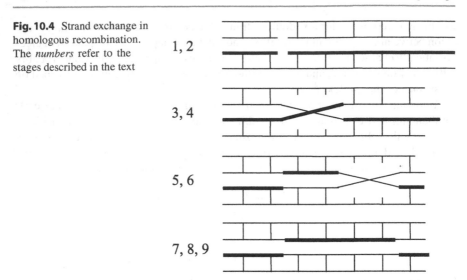

10.6.4 Summary of Sources of Genome Variation

Single-site mutations, common to all life-forms, may be due to mistakes in duplication (possibly caused by damage to the template base; e.g., due to ionizing radiation). A point mutation is a change in a single base (pair). Note that single insertions or deletions will change the reading frame; that is, all subsequent triplets will be mistranslated.

Microchromosomal and macrochromosomal rearrangements refer to large-scale changes involving many blocks of nucleotides. Tandem gene duplications may arise during DNA replication but, otherwise, the main source for chromosome rearrangement is meiosis.

Prokaryotes mostly do not reproduce sexually and, hence, do not undergo meiosis but, on the other hand, they are rather susceptible to "horizontal transfer" (i.e., the acquisition of genetic material from other bacteria, viruses etc.).[35]

The question of bias in single-site mutations is one of great relevance to evolution. The null hypothesis is that any mutation will occur with equal probability. If the mutation is functionally deleterious, according to the Darwinian principle it will not be fixed in the population, and the converse is true for functionally advantageous mutations. Kimura's "neutral" theory of evolution asserts that functionally neutral (i.e., neither advantageous nor deleterious) mutations will also become incorporated into the genome (leading to the phenomenon of "genetic drift").

A similar, but even more intriguing, question can be posed regarding bias in sites of chromosome breakage and crossover. At present, although it is recognized that the

---

[35]See Arber (1998).

likelihood of DNA duplication or moving is sequence-dependent, there is no overall understanding of the dependency.

10.7 Gene Expression

Gene expression refers to the processes (Fig. 10.1, d, e and f) whereby proteins are produced ("expressed") from a DNA template. It thus constitutes the bridge between genotype and phenotype. Whenever cells are not preparing for division (and many highly differentiated cells never divide), they are simply living, which means, in formal terms, that they are engaged in maintaining their essential variables within the domain corresponding to "alive" (Sects. 9.2 and 9.3). In certain environments, such as ocean floor sediments several kilometres thick, metabolic activity (of the bacteria that are presumed to be ubiquitous there) may be barely detectable (the degree of activity may be many orders of magnitude less than that of familiar laboratory bacteria, or that of those living parasitically inside a warm-blooded creature). Such environments are, moreover, unchanging or barely changing; hence, the vital processes can be maintained with very little need to change any of the parameters controlling them.

Most natural habitats show far more variety of conditions, however. Commonly encountered environmental disturbances include the fluctuating presence of toxic molecules and changes of temperature. Hence, cells need the ability to adapt (i.e., to modify their phenotypes to maintain their essential variables within the vital range). The formal framework for understanding this process was introduced in Chap. 9. Here we examine the molecular mechanisms of regulation that enable adaptation— the control of expression of different proteins as the cell proceeds round its cycle (Fig. 10.2) and as an organism develops (Sect. 10.8); development is a consequence of differential gene expression. The mechanism is essentially the same in all these cases. The entire process of gene expression is facilitated by many enzymes.

Despite the existence of elaborate machinery for regulating transcription (Sect. 10.7.2 ff.) Stochastic influences on expression and, hence, phenotype are discernible.[36]

10.7.1 Transcription

The essence of transcription is that RNA polymerases (RNAp, a large molecule with $M_r \sim 500\,000$) bind to certain initiation sites (sequences of DNA to which their affinity is superior) and synthesize RNA complementary to the DNA,[37] taking RNA monomers (nucleotide pyrophosphates) from the surrounding cytoplasm.

---

[36]Blake et al. (2003), Raser and O'Shea (2005).

[37]The transformation is given by: $\downarrow \dfrac{\text{A G C T}}{\text{U C G A}}$.

These enzymes catalyse the formation of a covalent bond between the nucleotide part of the monomer and the extant uncompleted RNA strand, and they release the pyrophosphate part into the cytoplasm as a free molecule. Presumably appropriate hydrogen bonds are formed to the DNA, RNA, and incoming nucleotide pyrophosphate, such that if the incoming nucleotide is correctly base-paired with the DNA template, it is held in the correct conformation for making a covalent bond to the extant RNA. The catalysis is reversible but is normally driven in the direction of RNA extension by a constant supply of monomers and the continual removal of the pyrophosphate.

Inition and termination of RNA synthesis are encoded within the DNA sequence. The RNAp is therefore similar in its action to the DNA polymerase in DNA replication.

The RNA folds up as it is synthesized (cf. Fig. 11.4), but extant structure may have to be disassembled as synthesis proceeds in order to achieve the final structure of the complete sequence.[38]

10.7.2 Regulation of Transcription

The key factor in transcriptional regulation is the affinity of RNAp for DNA. The prequisite for RNA production is the binding of RNAp in the initiation zone of the DNA. The binding affinity is *inter alia* influenced by the following:[39]

1. The binding of molecules to the RNAp;
2. The binding of molecules to the DNA initiation zone.

It is convenient to consider separately transcriptional regulation in prokaryotes and eukaryotes.

10.7.3 Prokaryotic Transcriptional Regulation

The main problem to be solved in prokaryotes is that different genes need to be active under different external conditions and during successive processes in the cell cycle. The primary control mechanism is via promoter sites situated upstream of that part of the DNA that will ultimately be translated into protein (cf. Fig. 10.3). For genes that need to be essentially constantly transcribed (the so-called housekeeping genes; i.e., those coding for proteins that are constantly required, such as those assembling the RNAp complex), there is no hindrance to RNAp binding to the inition zone

---

[38]See Fernández (1989).

[39]Suppression of transcription is not perfect. There appears to be a basal rate of transcription of some genes even in tissues in which they are not required. See Chelly et al. (1989) and Sarkar and Sommer (1989).

and beginning its work; only in exceptional circumstances might it be necessary to arrest production, whereupon a protein (called a repressor) will bind to a sequence within the inition zone (often immediately preceding the protein coding sequence) called the promoter, preventing the RNAp from binding to the DNA (Sauvageot's principle). Sometimes the transcription factor is simply the gene product. Conversely, for proteins seldom required, such as an enzyme for detoxifying a rarely encountered environmental hazard, the appropriate RNAp will normally have no affinity for the initiation zone, but should the toxin penetrate the cell, it will trigger the binding of a promoting (rather than inhibiting) transcriptional factor (called an activator) to the promoter site, whereupon the RNAp can bind and start its work.

Sometimes the translation of several (functionally related) genes is controlled by a single promoter. These structures of genes and promoter are called operons.

10.7.4 Eukaryotic Transcriptional Regulation

The requirements for gene regulation in eukaryotes are more complex, not least because, in a multicellular organism, as it differentiates many genes need to be permanently inactivated. Eukaryotes therefore have much richer possibilities for regulating transcription than prokaryotes. The mechanisms fall into five categories:

1. DNA methylation;
2. Chromatin conformation;
3. Binding of complementary ("antisense") RNA to key sites on the DNA;
4. Promoter sites and transcription factors (activators and repressors) as in prokaryotes;[40]
5. Competition for transcription factors by promoter sites on pseudogenes.

DNA Methylation
The enzymatic addition of methyl groups to cytosines prevents the gene from being transcribed. This inactivation can be reversed (demethylation), but some genes are irreversibly (permanently) inactivated (e.g., in the course of development), for example, by destruction of the start site. It is not well understood how these different degrees of inactivation come about. The interrelationship between histone modification (Sect. 10.4.4) and DNA methylation may well play a rôle.

Methylation—of 5′-C-G-3′ pairs (CpG, see Fig. 11.3)—is considered to be the major epigenetic mechanism at the molecular level.[41] The actual pattern of methylation is highly specific according to the cell type. In 98% of the human genome,

---

[40]Whereas a single RNAp operates in prokaryotes, there are at least three distinct ones in eukaryotes, accompanied by a host of "general transcription factors," which considerably increases the possible combinations of regulatory agents.

[41]The conventional view is that mammalian methylation occurs exclusively, or at least predominantly, at CpG pairs, but see, e.g., Doerfler et al. (1990) and Guo et al. (2014).

CpGs occur roughly once per 80 base pairs but, in the remainder, one finds CpG "islands"—sequences ranging from a few hundred to several thousand base pairs with a roughly fivefold abundance of CpGs. These islands almost always encompass gene promoters or exons; about half of all genes seem to contain such an island. CpGs within islands are normally unmethylated, whereas most of those without the islands are methylated (and hence transcriptionally inactive).[42] Methylation is a way of retaining information (gathered by the organism from its environment and from its own functioning) at the ontogenic level.

Chromatin Conformation and Modification

Long regarded as passive structural elements (despite the fact that the chromosome was known to undergo striking changes in compaction during mitosis), the histones (Sect. 10.4.4) are now perceived as actively participating in the regulation of gene expression. The essential principle is that the histones can be modified and unmodified by the covalent attachment and detachment of chemical groups, especially to and from the protein "tails" that protrude from the more compact core of the nucleosome. These result in changes in the protein conformation, affecting the conformation of the DNA associated with the histone and affecting the affinity and accessibility to RNAp. Acetyl groups have attracted particular attention, but methyl and phosphate groups and even other proteins also appear to be involved. The effect of these modifications is to control whether the associated gene is expressed. The modifications are catalysed by enzymes.

Currently, there are several ambiguities in the perception of nucleosome-modified gene expression regulation; for example, either acetylation or deacetylation may be required for enabling transcription and the modification can be local or global (affecting an entire chromosome). Are the effects of the modifications on the ability of transcription enzymes to bind and function at the DNA dependent on the modification of DNA shape, or rigidity, by the modified histones? There may also be proteins other than histones, likewise susceptible to modification, associated with nucleosomes. It is appropriate to consider the nucleus as a highly dynamic object full of proteins reacting with and diffusing to, from, and along the DNA.

RNA Interference

For many years, the rôles of RNA were thought to be confined to messenger RNA, transfer RNA, and ribosomal RNA; remarkably, the very extensive activity of the so-called "noncoding RNA" transcribed from intergenic regions and possibly introns in regulating gene expression was unsuspected until recently. Currently, two classes of this small (about two dozen nucleotides) RNA are recognized: microRNA (μRNA or miRNA) and small interfering RNA (siRNA). They appear to originate from their own microgenes, or are formed from RNA hairpins (cf. Fig. 11.5) resulting from mistranscribed DNA.

---

[42]Useful references for this section are Doerfler et al. (1990), Ramsahoye et al. (1665) and Bird (2002).

These small RNA molecules seem to be as abundant as mRNA, and their basic function is to block transcription by binding to complementary DNA sequences, or to block translation by binding to complementary RNA sequences. The varied applications of this function include plant defence against viruses.[43]

Promoter Sites and Transcription Factors

The affinity of RNAp to DNA is strongly dependent on the presence or absence of other proteins on the DNA upstream of the sequence to be transcribed (cf. Fig. 10.3), and associated with the RNAp. The principle of activation and repression by the binding of transcription factors to promoter sites is essentially as in prokaryotes; in eukaryotes, more proteins tend to be involved, allowing very fine tuning of expression.

Some molecules can directly interact with mRNA, altering its conformation and preventing translation into protein. This ability can be used to construct a simple feedback control mechanism; that is, the mRNA binds to its translated protein equivalent. mRNAs able to act in this way are known as riboswitches.

10.7.5 mRNA Processing

Post-transcriptional modification, or RNA processing, refers to the process whereby the freshly synthesized RNA is prepared for translation into protein. In prokaryotes, translation often starts while the RNA is still being synthesized; in eukaryotes, there is an elaborate sequence of reactions preceding translation. In summary, they are: capping; $3'$-polyadenylation; splicing; and export. Moreover, the whole process is under molecular surveillance and any erroneously processed RNA is degraded back into monomers.

Splicing is needed due to the introns interspersed in the DNA coding for protein. The initially transcribed RNA is a faithful replica of both introns and exons. This pre-mRNA is then edited and spliced (by the spliceosome, which is constituted from small nuclear riboprotein particles (snRNPs), each incorporating five small nuclear RNAs and several proteins bound to them). The DNA and the enzymes for transcription and post-transcriptional modification are enclosed in the lipid bilayer-based nuclear envelope, from which the edited RNA is exported (as messenger RNA, mRNA) into the cytoplasm for translation.

Alternative splicing of pre-mRNA is a powerful way of generating variant proteins from the same stretch of DNA; a majority of eukaryotic genes are probably processed in this way and, hence, the number of different proteins potentially available far exceeds the number of genes identified from the sequence of the genome. This method of generating variety is especially prominent in the generation of B-cell diversity in the immune system (Sect. 10.5).

---

[43] Voinnet (2001); Ding and Voinnet (2014).

10.7.6 Translation

The mature mRNA emerges from the nucle(ol)us where it is processed by the ribosomes, which are large ($M_r \sim 3 \times 10^6$ in bacteria; eukaryotic ones are larger), abundant (about 15 000 in an *E. coli* cell) protein-RNA complexes. In eukaryotes, ribosomes are typically associated with the endoplasmic reticulum, an extensive internal membrane of the cell. The overall process comprises initiation (at the start codon), elongation and termination (when the stop codon is reached). Elongation has two phases: In the first (decoding) phase, a codon of the mRNA is matched with its cognate tRNA carrying the corresponding amino acid, which is then added to the growing polypeptide; in the second phase, the mRNA and the tRNA are translocated one codon to make room for the next tRNA. As established by Crick et al. (1961) the mRNA is decoded sequentially in nonoverlapping groups of three nucleotides.[44] A messenger RNA may be used several times before it is degraded.

Some of the synthesized proteins are used internally by the cell; for example, as enzymes to metabolize food and degrade toxins and to build up structural components within the cell, such as lipid membranes and cytoskeletal filaments, and organelles such as the chloroplast. Other proteins are secreted to fulfil extracellular functions such as matrix building (for supporting tissue; or for biofilm) and other specialized functions, which become more and more complicated as the organism becomes more and more sophisticated. Another group of proteins modulate transcriptional, translational, and enzymatic activities. Many proteins have a dual function as a regulator and as something else—for example, an enzyme may also be able to modulate transcription, either of its own RNA or that of another protein.

It is estimated that about a third of newly synthesized proteins are immediately degraded by proteasomes, because they have recognizable folding errors.

10.8   Ontogeny (Development)

A multicellular organism begins life as a zygote, which is the diploid result of the union of two (haploid) gametes, male and female. The zygote then undergoes a series of divisions, the number of cells doubling each time; when 16 cells are present the zygote has developed into a morula. Its cells then compactify to form a two-dimensional shell (the blastoderm) enclosing a cavity (the blastocoele) filled with fluid, the overall object being called a blastocyst or blastula. The presence of maternal transcription factors regulates the initial pattern of gene activation. Rich possibilities ensue once several cells are formed, for they can emit and receive substances that activate or inhibit internal processes (including the ability to emit and receive these substances). At this stage the developing embryo can be modelled as a two-dimensional cellular automaton. The blastula then invaginates into two or three

---

[44]See Table 3.1 for the nucleic acid to amino acid transformation.

layers of cells, the ectoderm on the outside and the endoderm on the inside, with possibly a mesoderm between them (see also footnote a to Table 10.6). This object is called the gastrula, which may be modelled as a three-dimensional cellular automaton. The ectoderm forms the epidermis and the nervous system; the mesoderm forms bone, cartilage, muscle, blood etc.; and the endoderm forms the epithelium of the digestive and respiratory systems and their organs such as the liver.

The word "evolution" was originally coined to describe the unfolding of form and function from a single-celled zygote to a multicelled adult organism ("normal development"). Since it happens daily and can be observed in the laboratory, it is far more amenable to detailed scientific study than evolution comprising speciation and extinction over geological timescales.

The notion of evolution as the *unfolding* of parts believed to be *already existent in compact form* had already been formalized in 1764 by Bonnet under the name of preformation, and had been given a rather mechanical interpretation (i.e., unfolding of a highly compact homunculus produced the adult form).

Later, the term (evolution) came to be used to signify the epigenetic aspects of development. Epigenesis became the alternative to preformation, with the connotation of "order out of chaos." Both preformation and epigenesis contained the notion of coded instructions, but in the latter, at the time of its formulation the actual mechanism was conceived rather vaguely (e.g., by suggesting the cooperation of "inner and outer forces"). Nevertheless, it was firmly rooted in the notion of entelechy; in other words, the emphasis was on the potential for development, not on a deterministic path, which is entirely compatible with the cellular automaton interpretation of development. One might also refer to the interaction of genes with their environment.[45] "Environment" includes constraints set by the physical chemistry of matter in general. Wilhelm His clearly perceived the importance of general mechanical considerations in constraining morphology.

The term "ontogeny" was coined by Ernst Haeckel to signify the developmental history of an individual, as opposed to "phylogeny," signifying the evolution of a type of animal or plant (i.e., the developmental history of an abstract, genealogical individual).

It has been an important guiding principle that ontogeny is a synopsis of phylogeny. Very extensive observations of developing embryos in the eighteenth and nineteenth centuries led to a number of important empirical generalizations, such as von Baer's laws of development (e.g., "special features appear after the general ones"). It was clear that development embodied different categories of processes with different timescales largely uncoupled from one another: simple growing (the

---

[45]This is a very basic notion that crops up throughout biology. At present, there is no satisfactory universal formulation, however, but many interesting models have been proposed and investigated, including those of Érdi and Barna (1984) for neurogenesis, and Luthi et al. (1998) for neurogenesis in *Drosophila*. All of these models reduce to the basic formulation for the regulator (Sect. 9.4), discussed by Ashby (1956).

Table 10.5 Summary of ontogenetic paths (see text for further explanation)

Rate		Effect	Morphological result	Name
Soma	Gonads			
Fast	–	Acceleration	Recapitulation	Acceleration
–	Fast	Truncation	Paedomorphosis	Progenesis
Slow	–	Retardation	Paedomorphosis	Neoteny
–	Slow	Prolongation	Recapitulation	Hypermorphosis

isometric increase of size); growing up (allometric increase,[46] especially important in development of the embryo); and growing older (maturation). By adjusting these timescales relative to each other (heterochrony), different forms could be created.

Much debate has centred around *neoteny*—the retention of juvenile features in the adult animal (paedomorphosis)—and *progenesis*—the truncation of ontogeny by precocious sexual maturation. They can be thought of as, respectively, retardation and acceleration of development. If organ size (y) is plotted against body size (x) and standard shape is defined as $(y/x)_C$, retardation implies that this ratio occurs at larger x and acceleration occurs at smaller x. Another form of acceleration is "recapitulation"—previously adult features are pushed into progressively earlier stages of descendent ontogenies. Table 10.5 summarizes ontogenetic paths.

10.8.1 Stem Cells

Multicellular organisms begin life as a single cell, which divides, and the offspring, in turn, grow and divide and ultimately differentiate to create the variety of cells that constitute the organism's cellular repertoire. Stem cells may be defined as cells that can both self-renew (i.e., reproduce themselves) and differentiate into multiple cell types (lineages). The "ultimate" stem cell is totipotent and has the ability to form all cell types. In mammals, the fertilized egg, zygote, and the cells from the first four divisions (up to 16 blastomeres) are totipotent. Note, however, that strictly speaking these cells cannot self-renew (e.g., a zygote cannot divide to make two zygotes), and hence should not perhaps be called stem cells. Pluripotent stem cells are able to differentiate into the three fundamental types of embryonic germ layer, namely ectoderm, mesoderm, and endoderm (see footnote a to Table 10.6 for more explanation), from which all the more specialized cell types are derived. Lower down in the hierarchy are multipotent stem cells, which can form a small number of more specialized cells derived from a particular germ layer and constituting the somatic tissues. Fully differentiated cells are typically unable to divide.

---

[46] Allometric relations are of the type $y = bx^a$, where a and b are constants. $a = 1$ corresponds to isometry.

Table 10.6 The major divisions (phyla) of animals

Phylum	Characteristic[a]	Examples
Porifera	No permanent tissue	Sponges
Coelenterata (cnidaria)	2 or 3 layers of cells	Nematode worms
Ctenophora	2 or 3 layers of cells	Comb jellies
Annelida	Mesoderm has a cavity	Earthworms
Arthropoda ($\sim \frac{4}{5}$ of all animal species)	Jointed limbs	Insects, crustaceans, arachnids
Mollusca	True coelom	Snails, octopus
Echinoderma	Urchin-skinned	Starfish
Chordata[b]	Backbone, skull	–

[a] Tissue appears with the coelenterata, initially as two layers of cells—an outer (ectoderm) and an inner (endoderm)—separated by a structureless jelly. In the more advanced exemplars, a third layer of cells, the mesoderm, replaces the jelly. These are the three primary so-called germ layers of cells, which further differentiate into more specialized organs. The main animal tissue types are epithelial, connective, muscle, and nervous. The topology of the coelenterata is that of a simple sack. The mesoderm cavity that appears with the annelida develops into the coelom of the mollusca (cf. the main plant tissue types: epidermal, vascular, ground (subdivided into parenchyma (responsible for photosynthesis (the mesophyll), storage, etc.), collenchyma (structural) sclerenchyma (structural, without protoplasm; i.e., fibrous); meristematic ground tissue is responsible for growth.

[b] The chordata (craniata) are subdivided into subphyla including the vertebrata, whose classes comprise the familiar agnatha (lampreys etc.), fish, amphibians, reptiles, birds and mammals

10.8.2 Epigenesis

The fundamental problem of differentiation is that all of the cells have the same complement of genes. How, then, can different types arise? Pluripotent stem cells can be made to differentiate into neurons, for example, by exposing them to retinoic acid (at a concentration exceeding a certain threshold). If the initially differentiated cells then secrete a substance that blocks their as yet undifferentiated neighbours from differentiating, a stable population of two cell types results.[47] Differentiation is thus seen to be a typical complex phenomenon (cf. Sect. 7.4). If all cells were at all times identical, then, of course, differentiation could never occur. Even if all are endowed with the same maternal substance that induces differentiation, however, provided that the quantity of the substance is small enough for appreciable fluctuations in its concentration to occur (among, say, the 16 blastomeres), then they will not differentiate simultaneously, and if those that do so first can then prevent their neighbours from doing so, segregation is assured. A great variety of specific molecular mechanisms is available for the realization of such processes.

---

[47] Luthi et al. (1998).

10.8.3 The Epigenetic Landscape

Waddington introduced the term "epigenetics" as the name for the study of "the causal interactions between genes and their products, which bring the phenotype into being",[48] and it is particularly associated with the ontogenic level of phenotype; that is, possibly stable and preferably heritable changes in gene expression and phenotype not requiring changes in the sequence of the four fundamental bases of DNA (Fig. 11.3) in the genome. Waddington is also credited with introducing the vivid imagery of the "epigenetic landscape" (Fig. 10.5); this represents the process of successive decision-making during cellular development, most decisions implying different ultimate outcomes of cell differentiation.[49]

10.8.4 r and K Selection

In an ecological void (i.e., a new environment empty of life), at least of the types we are considering, or a highly fluctuating environment, growth is limited only by the coefficient r in equation (7.5) (r-selection). This circumstance favours progenesis—rapid proliferation at the cost of sophistication; and slight acceleration of development (cf. Table 10.5) leads to a disproportionately greater increase in fecundity.

In an older, more complex ecosystem (with a high density of organisms and intense competition for resources), or a very stable environment, growth is limited by its carrying capacity—the coefficient K in Eq. (7.5) (K-selection). This circumstance favours neoteny. Development is stretched out to enable the development of more sophisticated forms. There is no pressure to be fecund; the young offspring have a very low fitness relative to other species. The most successful beings are likely to be old and wise. The K-selective régime is the scenario for classical progressive evolution, characterized by a primary rôle for increasingly specialized morphology in adaptation, a tendency for size to increase, and hypermorphosis (the phyletic extension of ontogeny beyond its ancestral termination) enabled by delayed maturation. This also applies to human beings, who in our currently K-limited environment will evolve as much as they can, within phylogenetic constraints, towards more sophisticated forms. In this way economic growth can continue. Nevertheless, the human population has grown to such an extent that resources are being used at such a rate that some of them might actually be used up, for all practical purposes, in the near future. Unless substitutes can be found, this implies a diminishing K and consequentially a diminishing population.

Both r- and K-selection lead to diminished flexibility: respectively, in progenesis, by structural simplification caused by the loss of adult genes; and by overspecialization.

---

[48]Goldberg et al. (2007).
[49]See Gilbert (1991) for a critique and Buss and Blackstone (1991) for an experimental exploration.

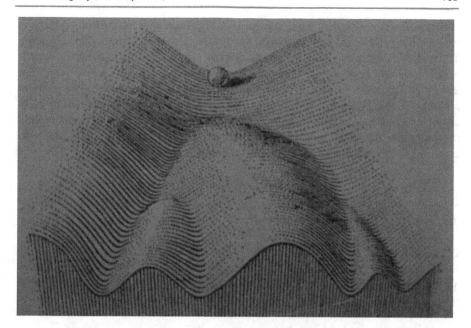

Fig. 10.5 Waddington's sketch of the epigenetic landscape (from C.H. Waddington, *The Strategy of the Genes: A Discussion of Some Aspects of Theoretical Biology*, London: George Allen and Unwin, 1957; reproduced with permission). Some explanation is given in C.H. Waddington, *Principles of Embryology*, New York: Macmillan, 1956: the spheroid represents a genotype and it has some bias (which in a physical realization of the model could be achieved by departures from sphericity, or an asymmetrical internal distribution of mass) corresponding to the particular initial conditions in some part of the newly fertilized egg. The surface slopes down towards the observer and at the saddle points (cf. Fig. 7.3 and the associated text) the genotype will move unpredictably to the *left* or to the *right*. The endpoint of the sequence of bifurcations will correspond to some typical organ. Waddington further proposed that the topography of the landscape, formed from a thin skin of some material, arose through a layer of genes beneath it, attached with guy-ropes to various points on the underside of the surface, the guy-ropes representing the "chemical tendencies which the genes produce"

A single species in a new, pristine environment simply proliferates until that niche is filled (r-selection). It also explores neighbouring genomes, and if these allow it to more successfully exploit some part of the environment (e.g., at the periphery of the zone colonized), a new species may result. Each new species itself makes the environment more complex, creating new niches for yet more species, and the environment is thereby transformed into one governed by K-selection.

10.8.5 Homeotic Genes

Homeotic genes regulate homeotic transformations; that is, they are involved in specifying body structures in organisms, homeosis (or homoeosis) being a shift in structural development. Homeotic genes encode a protein domain, the homeodomain, which binds to DNA and regulates mRNA synthesis; that is, it is a transcription factor. The part of the gene encoding the homeodomain is known as the homeobox, or *Hox* gene (in vertebrates). It is a highly conserved motif about 180 bases long. *Hox* and *Hox*-like genes (in invertebrates) are arranged consecutively along the genome and this order is projected onto, for example, the consecutive arrangement of body segments in an insect. Although considerable work has been done on elucidating the molecular details of homeotic transformations, it is not presently possible to encapsulate this knowledge in an algorithm for development.

10.9 Phylogeny and Evolution

Classical Darwinian theory[50] is founded on two observed facts:

1. There is (inheritable) variety among organisms.
2. Despite fecundity, populations remain roughly constant.

From these Darwin inferred that population pressure leads to the elimination of descendants less able to survive than slightly different congeners. Formally, therefore, evolution is a problem of selection. Only certain individuals (or species—see Sect. 10.9.1) are selected to survive. It is practically synonymous with natural selection, the "natural" being somewhat redundant.

Modern evolutionary theory is especially concerned with the following:

1. The levels at which change occurs (e.g., genes, cell lineages, individual organisms, species). Darwin dealt with individual organisms (microevolution); macroevolution deals with mass extinctions.
2. The mechanisms of change corresponding to the levels. The root of inheritable variation lies in the genes, of course; investigations of mechanisms operating at the higher levels subsume the lower-level mechanisms. The investigation of macroevolution has to deal with unusual (rare) events, such as the collision of Earth with a large meteor, and with avalanches of extinctions facilitated by trophic and other interactions between species.
3. The range of effects wrought by natural selection, and the timescales of change.

Critiques of classical Darwinism are legion. *Inter alia*, one may note the following: The selectionist explanation is always a construction *a posteriori*; evidence cited in favour of natural selection is often inconsistent; hence, rules are difficult to discern (examples: what is the selectionist advantage of the onerous migration of

---

[50]It is fairly well known that both Darwin and Wallace contributed independently; perhaps less well known is that priority appears to belong to Patrick Matthew (1831).

Comacchio eels to the Sargasso Sea for breeding? Why does the cow have multiple stomachs, whereas the horse (a vegetarian of comparable size) has only one? Why do some insects adopt marvellous mimickries allowing them to be concealed like a leaf, whereas others, such as the cabbage white butterfly, are both conspicuous and abundant?)—all one can say is that every surviving form must have been viable (i.e., of some selective advantage) or it would not have survived, and this is, of course, no proof that it is a product of selection; there appears to be no essential adaptive difference between specialization and nonspecialization—both are found in abundance; selection presupposes all of the other attributes of life, such as self-maintenance, adaptability, reproduction and so forth; hence, it is illogical to assert that these attributes are the result of selection; there is no evidence that progression from simple to complex organisms is correlated with better adaptation, selective advantage, or the production of more numerous offspring—adaptation is clearly possible at any level of organization, as evinced by the robust survival of very simple forms.

Although the classical theory ascribes competition between peers as a primordial motor of change, decisive evolutionary steps seem to have occurred when the relevant ecological niches were relatively empty, rather than in a period of intense competition.[51]

Arguments of this nature imply that the classical or orthodox view of evolution does not offer a satisfactory explanation of the observed facts. At present, we do not have one. It looks likely that principles of self-organization (Sect. 9.8), rooted in the same physicochemical laws governing the inanimate world, are involved. It would appear to be especially fruitful to focus on the constraints, on which a start has been made by Stephen Jay Gould (1977) with his picturesque image of spandrells in vaulted rooms: In well-known buildings, such as the San Marco cathedral in Venice, the decoration of the spandrells is a notable feature and contributes so significantly to the overall aesthetic effect that one's first impression is that they were designed into the structure by the architect. They are, however, an inevitable consequence of the vaulting and were used opportunistically for the decoration, much as feathers, developed to provide thermal insulation, seem to have been used opportunistically for flight—flight was an exaptation, not an adaptation. Other examples are now known at the molecular level, where existing enzymes are adapted to catalyse new, unrelated reactions.

The synthetic theory of evolution (sometimes called gradualism) asserts that speciation is a consequence of adaptation. Species are supposed to arise through the cumulative effects of natural selection acting on a background noise of myriads of micromutations. The genetic changes are not random (in contrast to classical natural selection), nor are they directed toward any goal. Change is opportunistic; that is, the most viable variants (in a given context) are selected. Selection takes place in vast populations. The sole mechanism is intraspecies microevolution.

---

[51] See Kirchner (2002) regarding limits on the rate of the filling process.

Fig. 10.6 Sketch of speciation according to the punctuated equilibrium concept

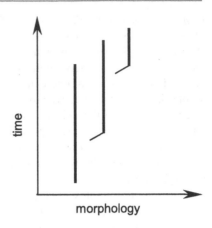

morphology

The synthetic theory is not in accord with the facts of palaeontology. Ruzhnetsev has emphasized that change is concentrated in speciation events. The time needed for a new species to become isolated seems to be negligible in paleontological (let alone geological) time: a few hundred years. Transitional forms are not observed (on the other hand, certain species have been stable for more than 100 million years). Speciation precedes adaptation. This theory is now usually called punctuated equilibrium (Fig. 10.6). It is in sharp contrast to gradualism, which predicts that the rate of evolution (i.e., the rate of speciation) is inversely proportional to generation time. There is little evidence for such a correlation, however. On the contrary, for example, the average species duration $\bar{D}$ for mammals is about 2 My.[52] Their initial Cenozoic divergence took place over about 12 My, but this would only allow time for about 6 speciations, whereas about 20 new orders, including bats and whales, appeared. Punctuated equilibrium interprets this as the rapid occupation (by speciation) of niches vacated by dinosaurs in the great mass extinction at the end of the Cretaceous era.

10.9.1 Group and Kin Selection

This topic arose through attempts to encompass altruism in evolutionary theory: an apparent paradox arises because the individual cost of altruism suggests that it should always be selected against (selection being considered to operate at the level of the individual).[53] The concepts of group selection and kin selection arose through attempts to incorporate the emergence of social behaviour into evolutionary theory. One posits a social structure in which individuals form clusters (cf. Sect. 8.3) or groups; a group exists if its members interact much more frequently with each

---

[52]See Stanley (1975) for a full discussion.
[53]McAndrew (2002).

other than with members of other groups. It is then asserted that natural selection operates at the level of the group. To avoid any implicit restriction to a single level of clustering, group selection is better referred to as multilevel selection. On the other hand, the concept of inclusive fitness (often referred to as kin selection) appears to deliver a similar result.

Let the *donor* be the executor of some altruistic act of kindness, and the *acceptor* the beneficiary. If $\mathfrak{R}$ is the genetic relatedness (given, for example, by Sewall Wright's coefficient of relationship) between donor and acceptor, $\mathfrak{B}$ the benefit (in terms of fitness) to the acceptor, and $\mathfrak{C}$ the fitness cost to the donor, then Hamilton's rule (drawing on the Price equation) states that altruism will be favoured if

$$\mathfrak{R}\mathfrak{B} > \mathfrak{C}. \tag{10.6}$$

The social insects (cf. Sect. 10.4.1) form a nice example of this rule in operation.

Actually group selection and kin selection are formally equivalent[54] and there seems to be little justification for the sometimes acrimonious disputes favouring one or the other mechanism.

Problem. Outline how Hamilton's rule suggests conditions under which cooperative behaviour can evolve.

10.9.2 Models of Evolution

Typical approaches assume a constant population of M individuals, each of whose inheritable characteristics are encoded in a string (the genome **s**, synonymous with genotype) of N symbols, $s_i, i = 1, \ldots, N$. N is fixed, and environmental conditions are supposedly fixed too. All of the individuals at generation t are replaced by their offspring at generation $t + 1$. The state of the population can be described by specifying the genomes of all individuals. Typically, values of M and N are chosen such that the occupancy numbers of most possible genomes are negligibly small; for example, if $N \sim 10^6$ and $M \sim 10^9$, $M \ll 2^N$, the number of possible genomes assuming binary symbols. In classical genetics, attention is focused on a few characteristic traits governed by a few alleles, each of which will be carried by a large number of individuals and each of which acts independently of the others (hence, "bean bag genetics"); modelling is able to take much better account of the epistatic interactions between different portions of the genome (which surely corresponds better to reality).

The model proceeds in three stages(cf. evolutionary computing, Sect. 8.1):

Reproduction: Each individual produces a certain number of offspring; the individual α at generation t is the offspring of an individual (the parent) that was living at generation $t - 1$ and which is chosen at random among the M individuals of the population.

---

[54]Marshall (2011).

Mutation: Each symbol is modified (flipped in the case of a binary code) at a rate μ; the rate is constant throughout each genome and is the same from generation to generation.

Selection: The genome is evaluated to determine its fitness $W(\mathbf{s}) = e^{F\mathbf{s}/C}$,[55] which, in turn, determines the number of offspring. C is the selective temperature.

The topography of a fitness landscape is obtained by associating a height $F(\mathbf{s})$ with each point $\mathbf{s}$ in genotype space. Various fitness landscapes have been studied in the literature; limiting cases are those lacking epistatic interations (i.e., interactions between genes) and those with very strong epistatic interations (one genotype has the highest fitness; the others are all the same). In the latter case the population may form a quasispecies (the term is due to Eigen, see footnote 33), consisting of close but not identical genomes. Distances between genomes s and s' are conveniently given by the Hamming distance:

$$d_{\mathrm{H}}(\mathbf{s}, \mathbf{s}') = \sum_{i=1}^{N} \frac{(s_i - s_i')^2}{4} , \qquad (10.7)$$

and the overlap between two genomes $\mathbf{s}$ and $\mathbf{s}'$ is given by the related parameter

$$\omega(\mathbf{s}, \mathbf{s}') = \frac{1}{N} \sum_{i=1}^{N} s_i s_i' = 1 - \frac{2 d_{\mathrm{H}}(\mathbf{s}, \mathbf{s}')}{N} . \qquad (10.8)$$

ω is an order parameter analogous to magnetization in a ferromagnet. If the mutation rate is higher than an error rate threshold, then the population is distributed uniformly over the whole genotype space ("wandering" régime) and the average overlap $\sim 1/N$ (see Sect. 10.6.2); below the threshold, the population lies a finite distance away from the fittest genotype and $\omega \sim 1 - \mathcal{O}(1/N)$.[56] Intermediate between these two cases (none and maximal epistatic interactions) are the rugged landcapes studied by Kauffman (1984).[57] More realistic models need to include changing fitness landscapes, resulting from interactions between species—competition (one species inhibits the increase of another), exploitation (A inhibits B but B stimulates A), or mutualism (one species stimulates the increase of another; i.e., coevolution).

As presented, the models deal with asexual reproduction. Sex introduces complications but can, in principle, be handled within the general framework.

These models concern microevolution (the evolving units are individuals); if the evolving units are species or larger units such as families, then one may speak of macroevolution. There has been particular interest in modelling mass extinctions, which may follow a power law (i.e., the number n of extinguished families $\sim n^\gamma$,

---

[55]The fitness of a phenotypic trait is defined as a quantity proportional to the average number of offspring produced by an individual with that trait, in an existing population. In the model, the fitness of a genotype $\mathbf{s}$ is proportional to the average number of offspring of an individual possessing that genotype.

[56]See Peliti (1996) for a comprehensive treatment.

[57]Cf. Sect. 7.2; see Jongeling (1996) for a critique.

with γ equal to about -2 according to current estimates). Bak and Sneppen (1993) invented a model for the macroevolution of biological units (such as species) in which each unit is assigned a fitness F, defined as the barrier height for mutation into another unit. At each iteration, the species with the lowest barrier is mutated—implying assigned a new fitness, chosen at random from a finite range of values. The mean fitness of the ecosystem rises inexorably to the maximum value, but if the species interact and a number of neighbours are also mutated, regardless of their fitnesses (this simulates the effect of, say, the extinction of a certain species of grass on the animals feeding exclusively on that grass[58]), the ecosystem evolves such that almost all species have fitnesses above a critical threshold; that is, the model shows self-organized criticality. Avalanches of mutations can be identified and their size follows a power law distribution, albeit with $\gamma \sim -1$. Hence, there have been various attempts to modify the model to bring the value of the exponent closer to the value (-2) believed to be characteristic of the Earth's prehistory.[59]

10.9.3 Further Remarks on Sources of Genome Variation

Non-Darwinian evolution ascribes the major rôle in molecular evolution to "genetic drift"—random ("neutral") changes in allele frequency (cf. Sect. 10.6.4). Classically, it is questionable whether genotypic differences without an effect on phenotype can affect fitness (in any sense relevant to evolution).[60] One should bear in mind that one of the engines of evolution, natural selection, operates on phenotype not genotype (to a first approximation at least) and, therefore, genes on their own are only the beginning of comprehending life; it is essential to understand how those genes are transformed into phenotype. To survive, however, a species or population needs adaptedness (to present conditions), (genetic) stability, and (the potential for) variability. Without stability, reproductive success would be compromised. Genetic variability is, of course, antithetical to stability, but phenotypic variability, reflecting control over which portion of the protein repertoire will be expressed, determines the range of environments in which the individual can survive and, hence, is equivalent to adaptedness to future conditions (cf. directive correlation and its degree, Sect. 9.2). The eukaryotic genome, with its resources of duplicate genes, pseudogenes, transposable elements, exon shuffling, polyploidy, and so forth, possesses the potential of phenotypic variability while retaining genetic stability. Prokaryotes lack these features, but they can readily acquire new genetic material from their peers or or from viruses (see footnote 35).

---

[58]For example the takahe feeds almost exclusively on snow grass.
[59]Newman (1996).
[60]See also Sect. 9.7

Table 10.7 The hierarchical scheme of the descriptive taxonomy of eukaryotes

Name	Example (1)	Example (2)
Kingdom	Animalia (metazoa)	Plantae (green plants)
Phylum	Chordata	Angiospermophyta
Subphylum	Vertebrata	–
Class	Mammalia	Monocotyledonae
Order	Primates	Asparagales
Suborder	Anthropoidae	–
Superfamily	Hominoidae	–
Family	Hominidae	Alliaceae
Genus	*Homo*	*Allium*
Species	*Sapiens*	*Sativum*
Individuals	Fred Bloggs	–

Examples are given for an individual human being and the culinary garlic

10.9.4 The Origin of Proteins

The random origin hypothesis[61] asserts that proteins originated by stochastic processes according to simple rules (i.e., that the earliest proteins were random heteropolymer sequences). This implies that their length distribution is a smoothly decaying function of length (determined by the probability that a stop codon will occur after a start codon has been encountered, in the case of templated synthesis without exons). On the other hand, the probability that a sequence can fold into a stable globular structure is a slowly increasing function of length up to about 200 amino acids, after which it remains roughly constant. Convolution of these two distributions results in a length distribution remarkably similar to those of extant proteins.

10.9.5 Taxonomy and Geological Eras

This section refers to the Tables 10.7 and 10.8 of the major groupings of living and growing things and the geological eras of the Earth.

Three lineages are recognized: the archaea (represented by extremophilic prokaryotes, formerly known as archaebacteria), the eubacteria (true bacteria, to which the mitochondria and chloroplasts are provisionally attributed), and the eukaryotes (possessing true nuclei). The eukaryotic kingdoms are animalia (metazoa), plantae, fungi, and protista (protozoa, single-celled organisms, including algae, diatoms, flagellates, amoebae, etc.). The approximate numbers of species of these

---

[61] See White (1994).

Table 10.8 History of the earth and earthly life

Name	Epoch[a]	Events	New or dominant life
–	4300	Earth formed	None
Phanerozoic	3500?	–	First life
	3000?	–	Stromatolites
	2500?	–	Mitochondria
	2000?	–	Bacteria
Palaeozoic			
Cambrian	570–500	–	Trilobites
Ordovician	500–440	–	–
Silurian	440–410	–	Fish, land (vascular) plants
Devonian	410–345	–	–
Carboniferous	345–280	Abundant plants	Giant insects, reptiles
Permian	280–225	Panguea (the single supercontinent), hot and dry; great mass extinction at the end	Reptiles
Mesozoic			
Triassic	225–190	Gondwanaland (the great southern continent)	–
Jurassic	190–134	Warm	Gymnosperms, ferns
Cretaceous	135–65	Mass extinction at end	Birds, dinosaurs
Cenozoic	**(tertiary)**		
Palaeocene	65–54	Volcanoes	Many
Eocene	54–38	Separation of eurasia	High diversity
Oligocene	38–26	Cooling	Low diversity
Miocene	26–7	Continental collisions	–
Pliocene	7–2.5	Himalayas, Alps	Elephants, *Australopicethus*
Cenozoic	**(quaternary)**		
Pleistocene	2.5–0.01	Last ice age	Woolly mammoth
Holocene	0.01–pres.	–	*H. sapiens*

[a]In millions of years before present

different kingdoms are currently estimated as 10^7 (metazoa), 2.5×10^5 (plantae), 2×10^5 (protozoa), and 5×10^4 (fungi).

Problem. Estimate the fraction of all possible DNA sequences that are represented in extant species.

References

Ahnert SE, Fink TMA, Zinovyev A (2008) How much non-coding DNA do eukaryotes require? J Theor Biol 252:587–592

Arber W (1998) Molecular mechanisms of biological evolution. In: Chou C-H, Shao K-T (eds) Frontiers in biology, pp 19–24. Taipei: Academia Sinica

Ashby WR (1956) An introduction to cybernetics. Chapman & Hall, London

Audit B, Audit N, Vaillant C et al (2002) Long-range correlations between DNA bending sites: relation to the structure and dynamics of nucleosomes. J Mol Biol 316:903–918

Bak P, Sneppen K (1993) Punctuated equilibrium and criticality in a simple model of evolution. Phys Rev Lett 71:4083–4086

Bird A (2002) DNA methylation patterns and epigenetic memory. Genes Dev. 16:6–21

Blake WJ, Kaern M, Cantor C, Collins JJ (2003) Noise in eukaryotic gene expression. Nature 422:633–637

Buss LW, Blackstone NW (1991) An experimental exploration of Waddington's epigenetic landscape. Phil Trans R Soc Lond B 332:49–58

Chan DC (2006) Mitochondrial fusion and fission in mammals. Annu Rev Cell Dev Biol 22:79–99

Chelly J, Concordet JP, Kaplan JC, Kahn A (1989) Illegitimate transcription: transcription of any gene in any cell type. Proc Natl Acad Sci USA 86:2617–2621

Crick FHC, Barnett L, Watts-Tobin RJ (1961) General nature of the genetic code for proteins. Nat Lond 192:1227–1232

Ding S-W, Voinnet O (2014) Antiviral RNA silencing in mammals: no newsis not good news. Cell Rep 9:795–797

Doerfler W, Toth M, Kochanek S, Achten S, Freisem-Rabien U, Behn-Krappa A, Orend G (1990) Eukaryotic DNA methylation: facts and problems. FEBS Lett 268:329–333

Eigen M (1976) Wie entsteht Information? Ber Bunsenges 76:1059–1081

Ellis RJ (2001) Macromolecular crowding: obvious but underappreciated. Trends Biochem Sci 26:597–604

Érdi P, Barna Gy (1984) Self-organizing mechanism for the formation of ordered neural mappings. Biol Cybern 51:93–101

Fernández A (1989) Pause sites and regulatory role of secondary structure in RNA replication. Biophys Chem 34:29–33

Gilbert SF (1991) Epigenetic landscaping: Waddington's use of cell fate bifurcation diagrams. Biol Philos 6:135–154

Goldberg AD, Allis CD, Bernstein E (2007) Epigenetics: a landscape takes shape. Cell 128:635–638

Gould SJ (1977) Ontogeny and phylogeny. Cambridge, Mass.: Belknap Press

Guo W, Chung W-Y, Qian M, Pellegrini M, Zhang MQ (2014) Characterizing the strand-specific distribution of non-CpG methylation in human pluripotent cells. Nucleic Acids Res 42:3009–3016

Hillman H (1991) The case for new paradigms in cell biology and in neurobiology. Edwin Mellen Press, Lewiston

Hittinger CT, Carroll SB (2007) Gene duplication and the adaptive evolution of a classic genetic switch. Nature 449:677–681

Hooke R (1665) Micrographia. The Royal Society, London

Jenuwein T, Allis CD (2001) Translating the histone code. Science 293:1074–1080

Jongeling TB (1996) Self-organization and competition in evolution: a conceptual problem in the use of fitness landscapes. J Theor Biol 178:369–373

Karlin S, Brendel V (1993) Patchiness and correlations in DNA sequences. Science 259:677–679

Kauffman SA (1984) Emergent properties in random complex automata. PhysicaD 10:145–156

Kellenberger E (1972) Assembly in biological systems. CIBA Found Symp (New Ser) 7:189–206

Kellermayer M, Ludány A, Jobst K, Szücs G, Trombitas K, Hazlewood CF (1986) Cocompartmentation of proteins and K+ within the living cell. Proc Natl Acad Sci USA 83:1011–1015

Kempner ES, Miller JH (1968) The molecular biology of *Euglena gracilis* IV. Cellular stratification by centrifuging. Exp Cell Res 51(1968) 141–149; idem, V. Enzyme localization. Exp Cell Res 51(1968) 150–156

Kim PM, Lu LJ, Xia Y, Gerstein MB (2006) Relating three-dimensional structures to protein networks provides evolutionary insights. Science 314:1938–1941

Kirchner JW (2002) Evolutionary speed limits inferred from the fossil record. Nature 415:65–68

Kornyshev AA, Leikin S (2001) Sequence recognition in the pairing of DNA duplexes. Phys Rev Lett 86:3666–3669

Lechler T, Fuchs E (2005) Asymmetric cell divisions promote stratification and differentiation of mammalian skin. Nature 437:208–275

Luthi PO, Preiss A, Chopard B, Ramsden JJ (1998) A cellular automatonmodel for neurogenesis in Drosophila. Physica D 118:151–160

Marshall JAR (2011) Group selection and kin selection: formally equivalent approaches. Trends Ecol Evol 26:325–332

Matthew P (1831) On naval timber and arboriculture. Adam Black, Edinburgh

McAndrew FT (2002) New evolutionary perspectives on altruism: multilevel-selection and costly-signaling theories. Current Directions Psychol Sci 11:79–82

McClintock B (1950) The origin and behavior of mutable loci in maize. Proc Natl Acad Sci USA 36:344–355

Newman MEJ (1996) Self-organized criticality, evolution and the fossil extinction record. Proc R Soc Lond B 263:1605–1610

Peliti L (1996) Fitness landscapes and evolution. In: Riste T, Sherrington D (eds) Physics of Bio-materials, pp 287–308. Kluwer, Dordrecht

Percus JK, Percus OE, Perelson AS (1993) Predicting the size of the T-cell receptor and antibody combining region from consideration of efficient self-nonself discrimination. Proc Natl Acad Sci USA 90:1691–1695

Ramsahoye BH, Biniszkiewicz D, Lyko F, Clark V, Bird AP, Jaenisch R (2000) Non-CpH methylation is prevalent in embryonic stem cells and may be mediated by DNA methyltransferase 3a. Proc Natl Acad Sci USA 97:5237–5242

Raser JM, O'Shea EK (2005) Noise in gene expression: origins, consequences, and control. Science 309:2010–2013

Rényi A (1953) Kémiai reakciók tárgyalása a sztochasztikus folyamatok elmélete segítségével. Magy Tud Akad Mat Kut Int Közl 2:83–101

Richards EJ, Elgin SCR (2002) Epigenetic codes for heterochromatin formation and silencing. Cell 108:489–500

Sander JD, Joung JK (2014) CRISPR-Cas systems for editing, regulatingand targeting genomes. Nat Biotechnol 32:347–355

Sarkar G, Sommer SS (1989) Access to a messenger RNA sequence or its protein product is not limited by tissue or species specificity. Science 244:331–334

Scherrer K, Jost J (2007) The gene and the genon concept. Mol Syst Biol 3(87):1–11

Shapiro JA (1992) Natural genetic engineering in evolution. Genetica 86:99–111

Solomon AK (1960) Red cell membrane structure and ion transport. J Gen Physiol 43:1–15

Stanley SM (1975) A theory of evolution above the species level. Proc Natl Acad Sci USA 72:646–650

Taft RJ, Pheasant M, Mattick JS (2007) The relationship between non-proton-coding DNA and eukaryotic complexity. BioEssays 29:288–299

Tonegawa S (1983) Somatic generation of antibody diversity. Nature (Lond) 302:575–581

Voinnet O (2001) RNA silencing as a plant immune system against viruses. Trends Genet 17:449–459

Voss RF (1992) Evolution of long-range fractal correlations and 1/f noise inDNA base sequences. Phys Rev Lett 68:3805–3808

Wakamoto Y, Ramsden JJ, Yasuda K (2005) Single-cell growth and division dynamics showing epigenetic correlations. Analyst, 311–317
Westermann B (2010) Mitochondrial fusion and fission in cell life and death. Nat Rev Mol Cell Biol 11:872–884
White SH (1994) Global statistics of protein sequences. A Rev Biophys Biomol Struct. 23:407–439

The Molecules of Life

11

11.1 Molecules and Supramolecular Structure

Table 11.1 gives some approximate values for the atomic composition of a cell. The atomic composition represents a highly reductionist view, somewhat akin to asserting that the informational content of *Macbeth* is $-\sum_{\text{alphabet}} p_i \log_2 p_i$, where p_i is the normalized frequency of occurrence of the ith letter of the alphabet. The next stage of complexity is to consider molecules (Table 11.2) and macromolecules (Table 11.3). This is still highly reductionist, however—it corresponds to calculating Shannon entropy from the vocabulary of *Macbeth*. Words are, however, grouped into sentences, which, in turn, are arranged into paragraphs. The cell is analogously highly structured—molecules are grouped into supramolecular complexes, which, in turn, are assembled into organelles. This structure, some of which is visible in the optical microscope, but which mostly needs the higher resolution of the electron microscope, is often called ultrastructure. It is difficult to quantify—that is, assign numerical parameters to it, with which different sets of observations can be compared. The human eye can readily perceive drastic changes in ultrastructure when a cell is subjected to external stress, but generally these changes have to be described in words.

The most prominent intracellular structural feature is the system of lipid bilayer membranes, such as the endoplasmic reticulum. Also prominent are the proteins such as actin, which form large filamentous structures constituting a kind of skeleton (the cytoskeleton). There are also many more or less compact (globular), large multiprotein complexes (e.g., the proteasome). Furthermore, proteins may be associated with lipid membranes or with the DNA. These structures are rather dynamic; that is, there is ceaseless assembly and disassembly, depending on the exigencies of survival. Some of them are described in more detail under the descriptions of the individual classes of molecules.

The interior of the cell is an exceedingly crowded milieu (compare the quantities of molecules with the dimensions given in Table 11.4). Although water constitutes about 70 % of a typical cell, very little of this water is free, bulk material. The very high

© Springer-Verlag London 2015
J. Ramsden, *Bioinformatics*, Computational Biology 21,
DOI 10.1007/978-1-4471-6702-0_11

Element	Rel. atomic fraction
H	100 000
C	5300
O	1600
N	1300
P	130
K, Na	80
S	40
Fe	5
Cu	1

Table 11.1 Atomic composition (selected elements) of a typical dried microbial cell

concentrations of molecules and macromolecules ensure that the cytoplasm is a highly viscous medium. Moreover, most of the macromolecules (e.g., proteins) are attached to larger structures such as the internal membranes. Kempner and Miller's

Table 11.2 Molecular composition of a typical microbial cell (the components are not uniformly dispersed in the cell)

Molecule	wt (%)	mol (%)	M_r[a]	No types	No molecules
DNA	1	–	3×10^9	1	1
RNA	6	–	(10^5)	500	250 000
Protein	15	–	5×10^4	1000	2×10^6
Saccharide	3	–	(10^4)	50	5000
Lipid[b]	2	0.1	10^3	40	2×10^7
Small[c]	2	1.0	10^2	500	10^7
Water	70	98.9	18	1	2×10^{10}

[a] Parentheses indicate approximate means of very broad ranges
[b] Including liposaccharides
[c] Metabolic intermediates, inorganic ions, and so forth

Table 11.3 Some characteristics of the macromolecules of a cell

Polymer	Monomer	Variety	Typical length	Bond variety[a]
DNA	Nucleotide[b]	4	2000	1
RNA	Nucleotide[b]	4	2000	1
Protein	Amino acid[c]	20	200	1
Polysaccharide	Monosaccharide	~10	20	~3

[a] That is, the type of bonding between monomers
[b] A nucleotide consists of a base, a sugar, and one or more phosphate groups. The variety resides solely in the bases
[c] An amino acid consists of a backbone part, identical for all except proline, and a side chain (residue) in which the variety resides

Table 11.4 Morphology and other properties of a typical eukaryotic cell (A typical prokaryote, such as the organism specified in Table 11.2, would have a diameter about 10 times smaller)

Property	
Shape	Sphere
Density	$1.025\,g/cm^3$
Radius	$5\,\mu m$
Volume	$5 \times 10^{-16}\,m^3$
Surface charge	$-10\,fC/\mu m^2$
Coat material	Polysaccharide
Coat thickness	$10\,nm$
Coat charge density	$-5\,MC/m^3$

classic experiments, in which they centrifuged intact cells to separate macromolecules from the water, demonstrated this very clearly—hardly any macromolecules were found in the aqueous fraction. This was in sharp contrast to the result of the traditional biochemical procedure of destroying all ultrastructure by mechanical homogenization, yielding an aqueous cytosol containing many dissolved enzymes (cf. Sect. 10.2.1).

The effect of the ultrastructure is twofold: to divide the cell up into compartments, not hermetically separated from one another but allowing access to different zones to be controlled, and to provide two-dimensional surfaces on which searching for and finding reaction partners is far more efficient than in an unstructured bulk.[1]

The separation of the macromolecules, which of course plays a crucial part in experimental bioinformatics, is dealt with in Part III.

11.2 Water

As seen from Table 11.2, water is overwhelmingly dominant in the cell. Water (H_2O) is a very unusual substance, as can be inferred from its extraordinarily high boiling point (compared with other molecules of comparable size) and large specific heat. A salient feature of the molecule is its great polarity—the bond between the oxygen and the hydrogen has a very strong ionic character. The electrostatic attraction between the positively charged hydrogen ($\delta+$) and the negatively charged electron lone pair on the oxygen ($\delta-$) constitutes the hydrogen bond (Fig. 11.1). It can be thought of as a redistribution of electron density from the covalent O–H bond to the zone between the H and the neighbouring O. This loss of electron density from the covalent O–H bond results in a weaker, more slowly vibrating bond.

---

[1] See Ramsden and Grätzel (1986).

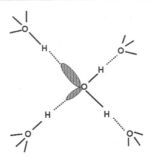

Fig. 11.1 A water molecule hydrogen-bonded to its congeners. The hydrogen atom is typically 0.10 nm from the oxygen to which it is covalently bonded (*solid lines*) and 0.18 nm from the neighbouring oxygen to which it is hydrogen-bonded (*dotted lines*). The energy of the hydrogen bond (H-bond) is about 0.1 eV (i.e., about $4k_B T$ at room temperature or about 2.4 kJ/mol)

Each water molecule can simultaneously accept and donate two hydrogen bonds (each hydrogen is a donor, and the oxygen bears two lone electron pairs). In flawless ice, the water molecules are H-bonded together in a tetrahedral arrangement.

The O–H infrared spectrum (of HOD in liquid D_2O) gives a very broad distribution of energies, implying a continuum from ice-like to nonbonding. In pure water at room temperature, about 10 % of the O–H groups and lone pairs (LP) are nonbonded; close to the boiling point, this percentage rises to about 40.

Bonded and nonbonded ions are in equilibrium:

$$H_2O_{\text{fully bonded}} \rightleftharpoons OH_{\text{free}} + LP_{\text{free}} \, , \qquad (11.1)$$

where the subscript "free" denotes nonbonded. LP_{free} and OH_{free} are, respectively, an electron donor (Lewis base) and electron acceptor (Lewis acid) and hence can interact with other species present in solution. An ion pair such as KCl interacts with both LP_{free} and OH_{free} in roughly equal measure; hence, KCl does not perturb the equilibrium (11.1), whereas (to take an extreme case) $NaB(C_6H_5)_4$ can only interact with LP_{free}, hence increasing the concentration of free OH groups. This kind of interaction has profound implications for macromolecular structure, as will be seen (Sect. 11.5).

11.3 DNA

Deoxyribonucleic acid is considered to be the ultimate repository of potentially meaningful information in the cell. DNA is poly(deoxyribonucleic acid), and the information is conveyed by the particular sequence of bases of the polymer. Each monomer unit has three parts: base, sugar, and phosphate (Fig. 11.2). The sugar (deoxyribose) and phosphate are always the same; the possibility of storing information arises through varying the base, for which there are four possibilities: the

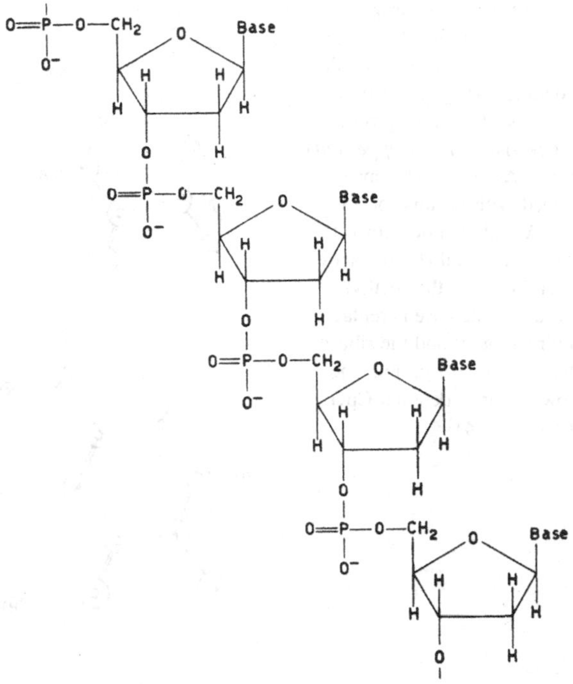

Fig. 11.2 Polymerized DNA.
The so-called 3' end is at the
upper left end, and the 5' end
is at the *lower right* (after
Ageno 1967; reproduced with
permission of the Accademia
dei Lincei)

purines adenine (A) and thymine (T), and the pyrimidines cytosine (C) and guanine
(G). The strand running from 5' to 3' is called the "sense" strand (i.e., it is used to
specify protein sequences via RNA), and the other one the "antisense" (antiparallel)
strand. Mainly only one strand encodes this information and the complementary one
serves to correct damage (Sect. 10.6.2).

Each base has the very important property of being able to H-bond with one
of the other three, the complementary base, significantly better than to any of the
others. This is perhaps the purest, most elementary example of molecular recognition.
Hence, a polymerized chain of monomers can serve as a template for the assembly of
a complementary strand. The purine pairs are linked by only two H-bonds, whereas
the pyrimidines are linked by three (Fig. 11.3). This means that the C–G base-pairing
melts (i.e., the H-bonds are broken) at a higher temperature than the A–T pairing.

As expected from their aromatic structure, the bases are planar. Figure 11.4 shows
the formation of the double helix. The genes of most organisms are formed by such
a double helix. The melting of the H-bonds as the temperature is raised is highly
cooperative (due to the repulsive electrostatic force between the charged phosphate
groups). On average, the separation into single stranded DNA occurs at about 80 °C
(at about 90 °C for sequences rich in C–G pairs, and at about 65 °C for sequences
rich in A–T pairs). These melting temperatures are lower at extremes of pH. Melting
leads to complete separation of the two chains, which is made use of in artificial gene

Fig. 11.3 The hydrogen-bonding patterns of complementary bases (thymine [T], adenine [A], guanine [G], cytosine [C], moving round clockwise from the *upper left*) (after Ageno 1967; reproduced with permission of the Accademia dei Lincei). In RNA, uracil (U) replaces thymine (i.e., the methyl group on the base is replaced by hydrogen) and the ribose has a hydroxyl group. The lower pair is denoted CpG (Sect. 10.7.4)

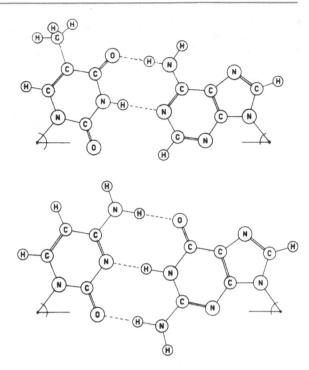

manipulation, as discussed in Part III. During *in vivo* replication, as discussed in the previous chapter, the chains are only separated locally.

Table 11.5 summarizes some significant discoveries relating to DNA.

It is now recognized that the structure, especially the sequence- and modific-ation-dependent rigidity (bending modulus) plays a profound rôle in the fidelity of replication, the regulation of transcription, and the movement of DNA through crowded milieux. The last aspect is of practical importance in DNA fractionation for sequencing, and so forth.

Under typical conditions of temperature, acidity, salt concentration, and so on prevailing in cells, the right-handed (Watson and Crick) double helix is the most stable structure, but others exist, such as the left-handed helix (Z-DNA), flips to which may play a rôle in gene activation. Circular DNA can be supercoiled; differing degrees of supercoiling affect the accessibility of the sequence to RNA polymerase and is thus a regulatory feature. There are several enzymes (topoisomerases, gyrases, and helicases) for changing DNA topology.

Double-stranded DNA is a rather rigid polymer, yet, despite its length, if stretched out in a straight line (about 1.2 mm for the DNA of *E. coli*), it is nevertheless packed into a cell only about 1 μm long. (Human DNA would be about 1 m long.)

A prominent feature of the DNA molecule is its high negative charge density due to the phosphate groups along the backbone. This gives DNA an ionic

Fig. 11.4 A stack of polymerized base pairs (*left*) distorted (*right*) by slightly twisting in order to form the double helix (after Ageno 1967; reproduced with permission of the Accademia dei Lincei)

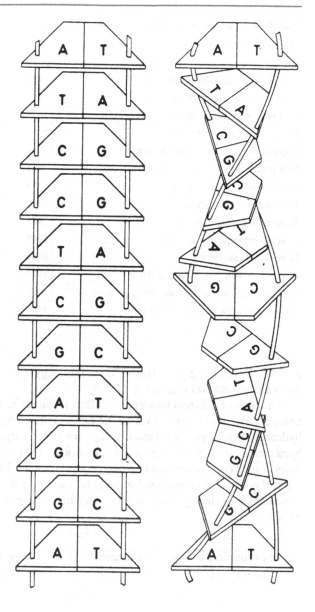

strength-dependent rigidity, which is also a significant factor affecting transcription and translation.

The rigidity can be quantified by the persistence length $\mathfrak{p}$, which depends on Young's modulus E:

$$\mathfrak{p} = E I_s / (k_B T) , \tag{11.2}$$

where I_s is the moment of inertia ($= \pi r^4 / 4$ for a cylinder of radius r), k_B is Boltzmann's constant, and T is the absolute temperature. For DNA, $r \approx 1.2\,\mathrm{nm}$ and

Table 11.5 Some milestones in molecular bioinformatics

Discovery or event	Year	Principal worker(s)
Nuclei contain an acidic substance	1869	Miescher
A tetranucleotide structure elucidated	1919	Levene
DNA identified as genetic material	1944	Avery
First protein (insulin) sequenced	1953	Sanger
DNA double helical structure	1953	Watson and Crick
Sequence hypothesis, central dogma	1957	Crick
First protein structure revealed (myoglobin)	1957	Kendrew, Perutz
Semiconservative replication	1958	Meselson and Stahl
DNA polymerase isolated	1959	A. Kornberg
Sequential reading of bases	1961	Crick
First protein sequence data bank	1965	–
Genetic code decrypted	1966	Crick
First protein structure data bank (PDB)	1971	–
First entire genome (*H. influenzae*) sequenced	1995	–
First multicellular genome (*C. elegans*)	1999	–

$E \approx 10^6$ N/m, giving $\mathfrak{p} \approx 60$ nm. The radius of gyration R_g of the polymer (length L) as a Gaussian coil is given by $(L\mathfrak{p}/6)^{1/2}$.

A mixture of different molecules of DNA is usually separated into its components using gel electrophoresis, in which the DNA is driven by an electric field through a hydrogel (usually polyacrylamide or agarose). Recently, model environments have been created from arrays of precisely positioned microfabricated pillars. Long polymers in such confined media move by reptation (rather like a snake moving through tall stiff grass—it is constrained laterally but can move along its length), in which they are confined to sliding along an imaginary tube between the pillars. The diffusivity D is, as usual,

$$D = k_B T / \delta , \qquad (11.3)$$

where δ is the drag coefficient and equal to $2\pi\eta L$, η being the viscosity of the solvent. The time for the polymer to diffuse out of its tube of length L is

$$\tau = L^2/(2D) \qquad (11.4)$$

but, in that interval, the polymer would have moved a distance equal to R_g if it had formed a Gaussian coil; the effective diffusion coefficient in the gel is then found from $D_{\text{gel}}/D = (R_g/L)^2$; hence,

$$D_{\text{gel}} = \frac{\mathfrak{p} k_B T}{12\pi\eta L^2} . \qquad (11.5)$$

Under the action of a relatively weak electric field and provided L is not too great, the mobility of the DNA in the gel is

$$\mu = \frac{\sigma \mathfrak{p}}{\sqrt{12\pi\eta L}} , \tag{11.6}$$

where σ is the charge per unit length of the DNA.[2]

Problem. On the basis of the above, devise a laboratory technique (including quantitative estimation of parameters for design and operation) for separating different molecules of DNA.

11.4 RNA

Ribonucleic acid, RNA, is rather similar to DNA. The most prominent difference is that the sugar is ribose rather than deoxyribose and that uracil rather than thymine is used as one of the two purine bases. These differences have considerable structural consequences. RNA does not occur as double helices; instead, base pairing is internal, forming parallel strands, loops ("hairpins"), and bulges (Fig. 11.5). It can therefore adopt very varied three-dimensional structures. It can pair (hybridize) with DNA.

RNA has five main functions: as a messenger (mRNA), acting as an intermediary in protein synthesis; as an enzyme (ribozymes); as part (about 60 % by weight, the rest being protein) of the ribosome (rRNA); as the carrier for transferring amino acids to the growing polypeptide chain synthesized at the ribosome (tRNA); and as a modulator of DNA[3] and mRNA interactions—small interfering RNA (siRNA; see Sect. 10.7.4).

Since ribozymes can catalyse their own cleavage, RNA can give rise to evolving systems; hence, it has been suggested that the earliest organisms used RNA rather than DNA as their primary information carrier. Indeed, some extant viruses do use RNA in that way.

A least-action approach—that is, minimizing the integral of the Lagrangian $\mathcal{L}$ (i.e., the difference between the kinetic and potential energies)—has been successfully applied to predicting RNA structure. The key step was finding an appropriate expression for $\mathcal{L}$. The concept can be illustrated by focusing on loop closure, considered to be the most important folding event. The potential energy is the enthalpy (i.e., the number n of contacts—here, base-pairings), and the entropy yields the kinetic

---

[2]For polymers confined by their congeners, a given chain can slowly escape from its tube by Brownian motion: The mobility μ of the whole chain N monomers long is μ_1/N, where μ_1 is the mobility of one monomer. Hence, from the Einstein relation $D_{tube} = \mu_1 k_B T/N$ and the relaxation time (to which viscosity is proportional) for tube length L ($\sim N$) to be lost and created anew, $\tau_{tube} \sim L^2/D = NL^2/(\mu_1 k_B T) \sim N^3$, in contrast to small molecules not undergoing reptation, for which $\tau \sim N$.

[3]Including heterochromatin formation.

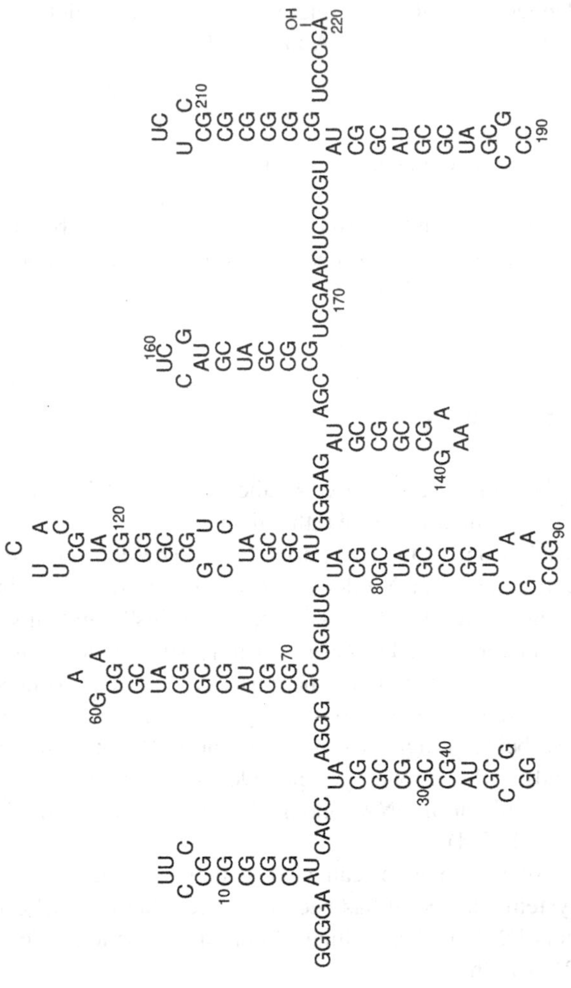

Fig. 11.5 A piece of RNA (from the $Q\beta$ replicase MDV-1) showing the characteristic loops formed by single-strand base-pairing

parameter. Folding is a succession of events in which at each stage as many new intramolecular contacts as possible are formed while minimizing the loss of conformational freedom (the principle of sequential minimization of entropy loss; SMEL). The entropy loss associated with loop closure is ΔS_{loop} (and the rate of loop closure $\sim \exp(\Delta S_{loop})$); the function to be minimized is therefore $\exp(-\Delta S_{loop}/R)/n$. A quantitative expression for ΔS_{loop} can be found by noting that the N monomers in an unstrained loop ($N \geq 4$) have essentially two possible conformations, pointing either inward or outward. For loops smaller than a critical size N_0, the inward ones are in an apolar environment, since the enclosed water no longer has bulk properties,[4]

---

[4]See Sinanoğlu (1981).

and the outward ones are in polar bulk water; hence the electrostatic charges on the ionized phosphate moieties of the bases will tend to point outward. For $N < N_0$, $\Delta S_{\text{loop}} = -RN \ln 2$, and for $N > N_0$, the Jacobson-Stockmayer approximation based on excluded volume yields $\Delta S_{\text{loop}} \sim R \ln N$. This allows $\mathcal{L}$ to be completely specified.[5]

11.5 Proteins

Proteins are appropriately named after the mythological being Proteus, who could assume many forms. The main functions of proteins are structural and catalytic. The catalytic function is especially important, for almost all of the other macromolecules of life, as well as small metabolites, are synthesized with the help of enzymes (catalysts). A rough overview of the protein world reveals the existence of the following:

Small polypeptides typically with no definite structure, acting as hormones, toxins, and so forth[6] (examples: bradykinin, mellitin);

Globular proteins typically able to assume a small number of stable configurations. This is the most numerous and varied class of proteins, comprising enzymes, transporters, regulators, motors, and so forth (examples: glucose oxidase, haemoglobin, kinesin, tumour necrosis factor α). Others in this class can polymerize to form fairly rigid rods (examples: flagellin, tubulin);

Fibrous proteins, which may be very long. They often have modular structures with many identical or at least very similar modules, which are folded up into small globules ("globulets") joined by short linker sections ("beads on a string").[7] Their rôle is mostly structural, both within and without the cell, but they actively interact with objects in their environment (e.g., neurites growing on them; i.e., as extracellular basement membranes they show chemical specificity) (examples: actin, collagen, laminin);

Glycoproteins, which may be very large, such that they form gels by entanglement. The polypeptide backbone is extensively decorated with relatively short polysaccharides. Typically they act as lubricants and engulfers (example: mucin);

Membrane proteins, which are also globular, but permanently embedded (tranversally) in a lipid bilayer membrane. They mainly function as channels, energy and signal transducers, and motors (examples: ATPase, bacteriorhodopsin, porin).

---

[5]See Fernández and Cendra (1996).
[6]See Zamyatnin et al. (2006).
[7]Rocco et al. (1987) describe this for fibronectin.

11.5.1 Amino Acids

The basic structure of an amino acid is $H_2N-C^{(\alpha)}HR-COOH$. At physiological pH, amino acids exist in zwitterionic form, $H_3N^+-C^{(\alpha)}HR-COO^-$. R denotes the variable side chain (residue); except for glycine (R=H), the $C^{(\alpha)}$ is asymmetric and hence chiral. The different residues are listed in Table 11.6.

Problem. Compare the abundances given in Table 11.6 with those predicted from Table 3.1, assuming that each nucleic acid triplet occurs with equal probability.

Amino acid polymerization takes place via elimination of water and the formation of the so-called peptide bond. Hence, a tripeptide with residues R_1, R_2, and R_3 has the structure $H_2N-C^{(\alpha)}HR_1-CO-N-C^{(\alpha)}HR_2-CO-N-C^{(\alpha)}HR_3-COOH$. Amino acids

Table 11.6 The natural amino acids in alphabetical order

Name	Three-letter abbreviations	One-letter code	Polarity[a]	Formula[b]	$\mathcal{A}$[c]
Alanine	ala	A	A	$-CH_3$	8.2
Arginine	arg	R	+	$-(CH_2)_3-NH-C(NH_2)_2^+$	3.9
Asparagine	asn	N	P	$-CH_2-CONH_2$	4.4
Aspartic acid	asp	D	−	$-CH_2-COO^-$	4.8
Cysteine	cys	C	P	$-CH_2-SH$	3.4
Glutamine	gln	Q	P	$-(CH_2)_2-CONH_2$	3.6
Glutamic acid	glu	E	−	$-(CH_2)_2-COO^-$	4.8
Glycine	gly	G	A	$-H$	7.6
Histidine	his	H	+	$-CH_2-[C_3N_2H_3]^+$	2.2
Isoleucine	ile	I	A	$-CH(CH_3)-CH_2-CH_3$	4.6
Leucine	leu	L	A	$-CH_2-CH(CH_3)_2$	7.3
Lysine	lys	K	+	$-(CH_2)_4-NH_3^+$	7.0
Methionine	met	M	A	$-(CH_2)_2-S-CH_3$	1.6
Phenylalanine	phe	F	A	$-CH_2-\phi$	3.5
Proline	pro	P	A	$-[C_3NH_7]$[d]	5.5
Serine	ser	S	P	$-CH_2-OH$	7.8
Threonine	thr	T	P	$-CH(OH)-CH_3$	6.5
Tryptophan	trp	W	A	$-CH_2-[C_8NH_6]$	1.2
Tyrosine	tyr	Y	P	$-CH_2-\phi-OH$	3.4
Valine	val	V	A	$-CH(CH_3)_2$	6.9

ϕ denotes a benzene ring. Square brackets denote a ring structure
[a] A apolar; P polar; + positively charged (at physiological pH); − negatively charged
[b] Of the side chain (residue)
[c] % abundance, from M.O. Dayhoff, ed., *Atlas of Protein Sequence and Structure*, Vol. 5. Washington DC: National Biomedical Research Foundation (1972)
[d] Incorporates the backbone $-NH_2$ in a ring structure

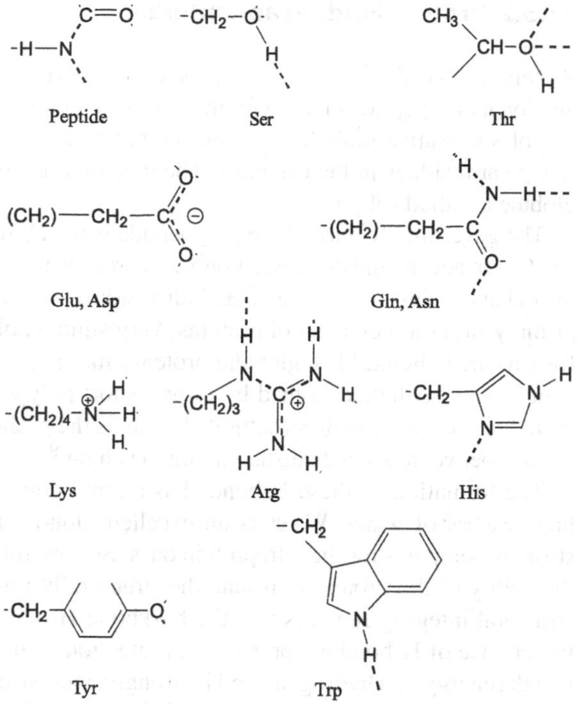

Fig. 11.6 Hydrogen-bonding capabilities of the peptide backbone and the polar residues (after Baker and Hubbard (1984)). Residues not shown are incapable of hydrogen bond formation

Peptide

Ser

Thr

Glu, Asp

Gln, Asn

Lys

Arg

His

Tyr

Trp

polymerized into a polypeptide chain are usually called peptides. The CO–N bond is in resonance with the C=O bond and is therefore rigid, the CO–N triatom system being planar; but the N–C$^{(\alpha)}$ and C$^{(\alpha)}$HR$_1$–CO bonds are free to rotate independently. Two dihedral angles, ϕ and ψ respectively, per amino acid therefore suffice to completely characterize the conformation of a polypeptide chain. A Ramachandran plot of ψ versus ϕ can be constructed for each amino acid showing the allowed conformations; constraints arise due to the overlaps between the atoms attached to the N–C$^{(\alpha)}$–C backbone.[8]

The amino acids can be classified in several ways according to their residues. A binary classification groups them as apolar (incapable of hydrogen bonding) or polar (see Fig. 11.6). The polar residues can be further classified into net hydrogen bond donors and acceptors. Other binary classifications are electrostatically charged (ionizable) and uncharged; big and small; and glycine or not.

---

[8]Another kind of Ramachandran plot is used to represent the structure of an entire polypeptide chain, by plotting the actual values of ψ versus ϕ in the folded structure of each amino acid.

11.5.2 Protein Folding and Interaction

Proteins are synthesized *in vivo* by the consecutive addition of amino acids to form an elongating peptide chain with the conformation of a random coil in the aqueous cytoplasm. Native globular proteins are compact stable structures with no or very few polar residues in their interior. The transition from a random coil to an ordered globule is called folding.

The governing feature of the polypeptide is the ability of the peptide unit –N–C–C(=O)– to accept and donate H-bonds. Geometrical constraints allow the ith residue in a chain to bond with the $(i \pm 3)$th residues to form the α-helix, which is the primary structural element of proteins. Very simple polypeptides (e.g., polyalanine) form a pure α-helix. Most globular proteins, made up of many different amino acids, contain short α-helices joined by turns—short polypeptide segments of no special structure. The other main structural element is the β-sheet, in which the H-bonds are formed between peptides distant along the chain.[9]

The formation of these H-bonds has to, and does, take place in the presence a huge excess of water. Water is an excellent donor and acceptor of H-bonds and strongly competes for the intraprotein ones. Successful folding therefore depends on the ability of the protein to isolate the structurally important H-bonds from water; structural integrity requires that the backbone H-bonds be kept dry. The energetic importance of H-bond wrapping (i.e., protection from water) can be seen by noting that the energy of a hydrogen bond is strongly context-dependent. In water, it is about 2 kJ/mol; *in vacuo*, it increases eightfold to tenfold. Wrapping will therefore greatly contribute to the enthalpic stabilization of globular protein conformation.

A poorly desolvated H-bond is called a dehydron.[10] The dehydron is under-wrapped and, therefore, overexposed to water (i.e., wet), because there are insufficient apolar groups in its vicinity. The only way for a protein to diminish the presence of water around a hydrogen bond is to bring apolar residues unable to form H-bonds with water into its vicinity; by keeping water away, hydrophobic groups, such as methyl and ethyl, are powerful intramolecular H-bond enhancers. The *dehydronic force* is thus a three-body force involving the H-bond donor, the H-bond acceptor, and the apolar residue. It is formally defined as the drag exerted by a dehydron on a test residue; that is,

$$F = -\nabla_{\mathbf{R}} \left(\frac{1}{4\pi\epsilon\mathbf{R}} \frac{qq'}{r_0} \right) , \tag{11.7}$$

where $\mathbf{R}$ is the position of the hydrophobic test residue measured perpendicularly from the H-bond, q and q' are the net charges, and r_0 is the O–H distance of the H-bond. Typically, F is about 7 pN at $\mathbf{R} = 6$ Å.

---

[9]As shown in Fig. 11.6, some residues can also participate in hydrogen bonding, but the backbone peptide H-bonds (or potential H-bond donors and acceptors) are of course more numerous and, hence, more significant.

[10]The dehydron concept is due to A. Fernández. See, for example, Fernández and Scott (2003) and Fernández et al. (2002, 2003).

The three-dimensional structure of a protein (as encoded in a pdb file) can be interrogated to reveal dehydrons. Hydrogen bonds are operationally defined as satisfying the criteria of an N–O distance of 2.5–3.5 Å and the angle between the NH and CO bonds equal to 45°. The dehydration domain of an H-bond is defined as two spheres of equal size centred on the $C^{(\alpha)}$s of the amino acids paired by the H-bond. The radius of the spheres (around 6.5–7 Å) is chosen to slightly exceed the typical distance between nonadjacent $C^{(\alpha)}$s; hence, the spheres necessarily interact. The extent of wrapping is given by the number ρ of hydrocarbon groups within the dehydration domains. A well-wrapped H-bond has $\rho = 15$; most soluble monomeric globular proteins have a ρ around this value, averaged over all the backbone H-bonds.

Wrapping defects are decisive determinants of protein–protein (and other) interactions. If the stable conformation of a globular protein is such that there are some unavoidably underwrapped H-bonds on its solvent-acessible surface, then that protein will be sticky; the underwrapped H-bonds will be hotbeds of stickiness.[11] Any other surface able to provide an appropriate arrangement of apolar groups will strongly bind to the dehydronic region (provided that geometric constraints—shape complementarity—are satisfied). The completion of the desolvation shell of a structure-determining H-bond has the same significance in understanding protein structure and interactions as completing electron shells has in understanding the periodic table of the elements in chemistry. Indeed, the dehydron concept is needed to computationally fold a peptide chain *ab initio*.

Examination of protein-protein interaction interfaces fully bears out the dehydron interpretation. Appropriate complementarity is achieved by overexposed apolar groups and dehydrons (rather than H-bond acceptors and donors, or positively and negatively ionized residues, although these may play a minor rôle). One also notes that each subunit of haemoglobin, a very stable and soluble (i.e., nonsticky) protein, has just three dehydrons: Two are at the interface with the other subunits, and one is the bond connecting residues 5 and 8 (i.e., flanking the sickle cell anaemia mutation site at residue 6). In contrast, the prion protein, which is pathologically sticky, has an extraordinarily high density of dehydrons (mean ρ is only about 11).

There are also evolutionary implications. It has long been realized that the evolution of proteins via mutations in their corresponding genes is highly constrained by the need to maintain the web of functional interactions. There is a general tendency for proteins in more evolved species to be able to participate in more interactions; they have more dehydrons. For example, mollusk myoglobin is a perfectly wrapped protein and functions as a loner. Whale myoglobin is in an intermediate position, and human myoglobin is poorly wrapped, hence sticky, and operates together with other proteins as a team. Although the folds in a protein of given function are conserved as species diverge, wrapping is not (even though the sequence homology might still be as much as 30%). Structural integrity becomes progressively more reliant on the interactive context as a species becomes more advanced.

---

[11]Empirically, a certain threshold density of dehydrons per unit area should be exceeded for a surface to qualify as sticky.

A corollary is that the proteins of more complex species are also more vulnerable to move into pathological states. The prion diseases form a good example; they are unknown in microbes and lower animals. Moreover, they mainly attack the brain, the most sophisticated and complex organ in the living world.

11.5.3 Experimental Techniques for Protein Structure Determination

High-throughput methodology (also called structural genomics) comprises the following steps:

1. Select the gene for the protein of interest.
2. Make the corresponding cDNA.
3. Insert the cDNA into an expression system.
4. Grow large volumes of the protein in culture (if necessary with appropriate isotopic labelling of C and N).[12]
5. Purify the protein (using affinity chromatography).
6. Crystallize the protein (often unusual salt conditions are required) and record the X-ray diffractogram,[13] or carry out nuclear magnetic resonance spectroscopy (one or more of ^{1}H, ^{13}C, ^{15}N) with a fairly concentrated solution of the protein to yield an adjacency matrix (cf. Sect. 7.2) from which the pattern of through-bond and through-space couplings can be derived.
7. Calculate the atomic coordinates.
8. Refine the structure by minimizing interatomic potentials, or use Ramachandran plots.

Under favourable conditions, X-ray diffraction and nuclear magnetic resonance spectroscopy (n.m.r.) can yield structures at a resolution of 1 Å. Some of the difficulties in these procedures are as follows:

1. The protein may not crystallize. Membrane proteins are especially problematical, but their structures may be obtainable from high-resolution electron diffraction of two-dimensional arrays, or by crystallizing them in a cubic-phase lipid.
2. Hydrogen atoms are insufficiently electron dense to be registered in the X-ray diffractogram (they are detectable in experimentally more onerous neutron diffraction).

---

[12]For some of the problems associated with the production of recombinant proteins, see Protein production and purification, Nature Methods 5 (2008) 135–146.

[13]Multiple isomorphous replacement—MIR—whereby a few heavy atoms are introduced into the protein, which is then remeasured, is used to determine the diffraction phases. The heavy atoms should not, of course, induce any changes in the protein structure.

3. Energy refinement will yield the majority structure. Most proteins have two or more stable structures, which may be present simultaneously, although in unequal proportions.
4. The crystal structure, or the structure in concentrated solution, may not be representative of the native structure(s).
5. Nuclear magnetic resonance cannot cope with large proteins (the spectra become too complicated, and the assignment of peaks to the individual amino acids along the sequence becomes problematical).
6. Nuclear magnetic resonance yields a set of distance constraints, but there are usually so many that the problem is overdetermined, and no physically possible structure can satisfy all of them.

Protein stability can be assessed by determining the structure of a protein at different temperatures. Since thermal denaturation is accompanied by a large change in specific heat, whose midpoint provides a quantitative parameter characterizing stability, microcalorimetry is a useful technique for assessing stability.

11.5.4 Protein Structure Overview

The techniques described in the previous subsection revealed that proteins have a compact structure akin to a ribbon folded back and forth. Drop a piece of thick string about a metre long on a table, pick it up, and push it together between one's hands. This gives a fair impression of typical protein structure at very low resolution. α-helices and β-sheets are called secondary structures (the primary structure is the sequence of amino acids). The arrangement of secondary structure elements is called the tertiary structure. Quaternary structure denotes arrangements of individual folded peptide chains (e.g., subunits) to form supramolecular complexes. Quinary structure is the network of other proteins with which a protein interacts.

The number of basic shapes in which proteins fold (i.e., the variety of tertiary structures) seems to be far smaller ($\sim 10^4$) than the number of possible sequences. Individual examples of sequences with less than 10 % homology folding into essentially the same structure are known. Some folds are very common, whereas others are rare.

11.6 Polysaccharides

Monosaccharides (sugars) are carbohydrates whose chemical composition is give by the empirical formula $(CH_2O)_n$, with typically $n = 3, 4, 5$, and 6. They are linked together via one of their oxygen atoms in an ether-like linkage to form oligomers and polymers. Saccharide monomers have many –OH groups, and there is much variety in their choice for linking. Some oligosaccharides are metabolic intermediates; they are very often used to modify proteins and lipids, with profound influence on their structure and reactivity.[14] For example, if one sugar is missing from transferrin, an iron-transporting protein in the blood with several glycosylated amino acids, the bearer has an abnormal skin colour, liver problems, and so forth. Oligosaccharides are extensively used to confer specificity of binding (e.g., in the immune system). Longer polysaccharides are used to store energy and as structural components. Their assembly is not templated but is accomplished by enzymes. There is considerable variety in the sequence of nominally identical heteroöligosaccharides.

Cellulose is a long unbranched chain of glucose monomers linked head to tail. As the major constituent of plant cell walls, cellulose is probably more abundant on Earth than any other organic material. The chains are packed side by side to form microfibrils, which are typically a mixture of two crystalline forms, I_α and I_β, and whose diameter ranges from about 3 nm in most plants to about 20 nm in sea squirts. The chains are held together by H-bonds.[15]

Problem.   Examine whether polysaccharides could be used as the primary information carrier in a cell.

11.7   Lipids

Lipids are not polymers, but in water they spontaneously assemble to form large supramolecular structures (planar bilayer membranes and closed bilayer shells, called vesicles). Lipids are amphiphiles; that is, they consist of a polar moiety (the "head") attached to an apolar one (the "tail," typically an alkane chain). The structures formed when lipids are added to water depend on the relative sizes of the polar and apolar moieties. If the tail is thinner than the head, as with many detergents, micelles, compact spherical aggregates with all the head facing outward, may form. Natural lipid molecules are typically roughly cylindrical—the head has about the same diameter as the tail—and readily form planar or slightly curved membranes (Fig. 11.7). Obconical shapes (head larger than tail) favour convex structures of small radius, such as endosomes or the borders of large (hydrophilic) pores in planar bilayer membranes. Conical shapes (such as phosphatidylethanolamine, which has a very small head) oppose this tendency.

A large number of natural lipids are known and found in natural membranes; both the head groups and tails can be varied. A small selection is shown in Fig. 11.8. The lipid repertoire of a cell or organism is called the "lipidome". This diversity allows the shape, fluidity, permeability, affinity for macromolecules, and so on of membranes to be adjusted. The biosynthesis of lipids and other membrane components such as cholesterol is, of course, carried out by enzymes, but the regulation of their

---

[14]See Dwek and Butters (2002) for an overview.
[15]See also "Symbols for specifying the confirmation of polysaccharide chains", Eur. J. Biochem. 131 (1983) 5–7, or Pure Appl. Chem. 55 (1983) 1269–1272.

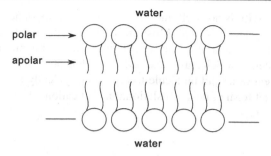

Fig. 11.7 A bilayer lipid membrane formed by two apposed sheets of molecules

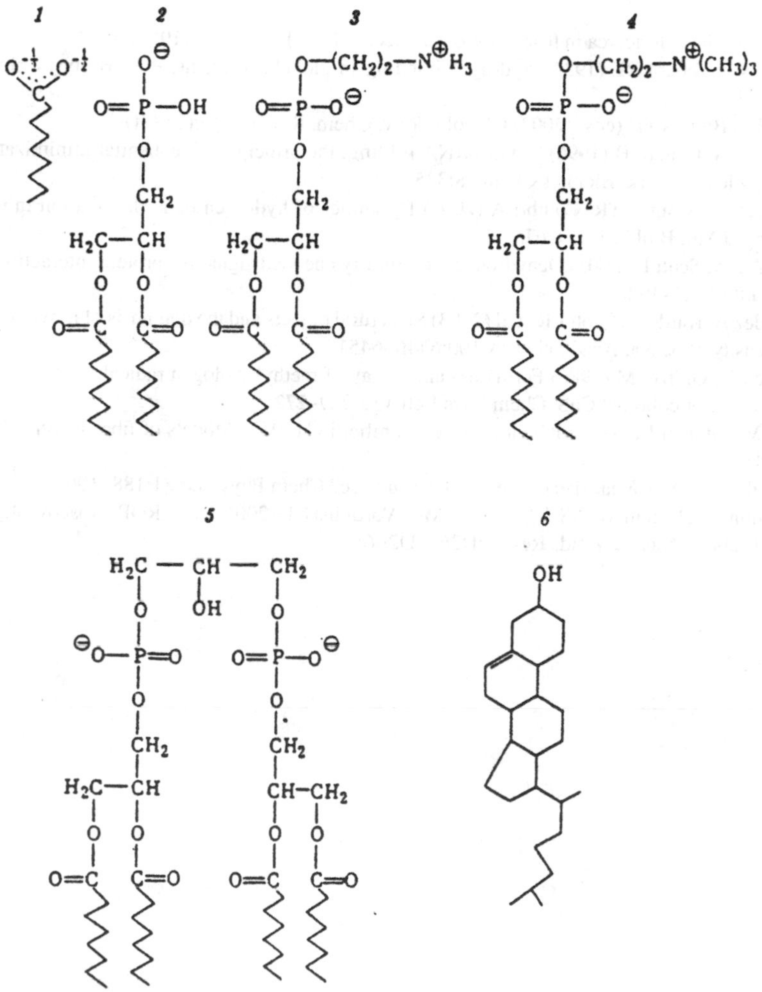

Fig. 11.8 Some naturally occurring lipids and membrane components. *1*, a fatty acid; *2*, phosphatidic acid; *3*, phosphatidylethanolamine; *4*, phosphatidylcholine; *5*, cardiolipin (diphosphatidylglycerol); *6*, cholesterol

abundance and activity is not well understood, and the importance of their variety has probably been underestimated. Most enzymes are attached to membranes and the lipids probably play a far more active rôle than merely functioning as a passive matrix for the protein—which may constitute more than 50 % of the membrane. The covalent attachment of a lipid molecule to a protein, typically at a terminal amino acid, is a significant form of post-translational modification.

References

M. Ageno, Linee di ricerca in fisica biologica. Accad. Naz. Lincei 102 (1967) 3–50

Baker EN, Hubbard RE (1984) Hydrogen bonding in globular proteins. Prog Biophys Mol Biol 44:97–179

Dwek RA, Butters TD (eds) (2002) Glycobiology. Chem. Rev. 102(2):(283 ff)

Fernández A, Cendra H (1996) In vitro RNA folding: the principle of sequential minimization of entropy loss at work. Biophys Chem 58:335–339

Fernández A, Sosnick TR, Colubri A (2002) Dynamics of hydrogen bond desolvation in protein folding. J Mol Biol 321:659–675

Fernández A, Scott R (2003) Dehydron: a structurally encoded signal for protein interaction. Biophys J 85:1914–1928

Fernández A, Kardos J, Scott LR et al (2003) Structural defects and the diagnosis of amyloidogenic propensity. Proc Nat Acad Sci USA 100:6446–6451

Ramsden JJ, Grätzel M (1986) Formation and decay of methyl viologen radical cation dimers on the surface of colloidal CdS. Chem Phys Lett 132:269–272

Rocco M, Infusini E, Daga MG, Gogioso L, Cuniberti C (1987) Models of fibronectin. EMBO J 6:2343–2349

Sinanoğlu O (1981) What size cluster is like a surface? Chem Phys Lett 81:188–190

Zamyatnin AA, Borchikov AS, Vladimirov MG, Voronina OL (2006) The EROP-Moscow oligopeptide database. Nucleic Acids Res 34:D261–D266

Part III
Applications

Introduction to Part III

12

Figure 12.1 is a simplified version of Fig. 10.1 that highlights the principle objects of investigation of bioinformatics. The field could be said to have begun with individual gene (and hence protein) sequences; typical problems addressed were the extraction of phylogenies from comparing sequences of the same protein over a wide range of different species and the identification of a gene of unknown function by comparison with the knowledge base of sequences of known function, via the inferential route:

$$\text{sequence homology} \Rightarrow \text{structural homology} \Rightarrow \text{functional homology.} \quad (12.1)$$

There are, however, plenty of examples of structurally similar proteins with different sequences or functionally different proteins with similar structures. Associated with these endeavours were technical problems of setting up and maintaining databases of sequences and structures.

The bioinformatics landscape was dramatically transformed by the availability of whole genomes and, at roughly the same time (although there was no especial connexion between the developments), whole proteomes and whole metabolomes. Far wider-ranging comparisons could now be carried out; in particular, a global vision of regulation seemed to be within grasp. Part III focuses on these developments; Table 12.1 recalls the magnitude, at the level of the raw materials, of the problems to be solved.

Genomics is concerned with the analysis of gene sequences, and there are two main territories of this work: (1) comparison of gene sequences, that is analysis of the relation of a given sequence with other sequences (external correlations); and (2) analysis of the succession of symbols in sequences (internal correlations). The first attempts to elucidate the function of sequences whose function is unknown by comparing the "unknown" sequence with sequences of known function. It is based on the principles that similar sequences encode similar protein structures, and similar structures encode similar functions (there are, however, many examples for which these principles do not hold). One also compares sequences known to code for the same protein (functionally speaking) in different organisms, in order to deduce phylogenetic relationships. A further branch of this territory compares the sequences of healthy and diseased organisms, in an attempt to assign genetic causes to disease. The second territory attempts to find genes (and, ultimately, other

© Springer-Verlag London 2015

J. Ramsden, *Bioinformatics*, Computational Biology 21,

DOI 10.1007/978-1-4471-6702-0_12

Fig. 12.1 The relation among genes, mRNA, proteins and metabolites. The *curved arrows* in the upper half of the diagram denote regulatory processes

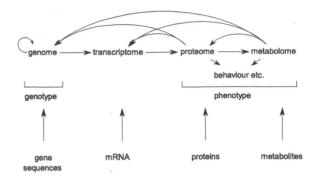

Table 12.1 Approximate numbers (variety) of different objects in the human body

Object	Number
Genes	30 000
mRNA	10^5
Proteins[a]	3×10^5
Expressed proteins[b]	10^3-10^4
Cell types	220
Cells[c]	$10^{13}-10^{14}$

[a] Potential repertoire
[b] In a given cell type
[c] Excluding microbial cells hosted within the body and which may be comparably numerous

functionally important sequences such as those involved in regulation) via linguistic inhomogeneities and to assign function to the genes by searching for regularities (the "grammar" of the sequence). In its purest form, genomics could be viewed simply as the study of the nonrandomness of DNA sequences. This endeavour is still inchoate, since the regularities and their relation to function are not understood. One may, however, be able to predict the structure from the sequence, which can then be used to advance the search for function. Even coarse indications may be useful; for example, transmembrane proteins typically possess several α-helices, traversing the lipid bilayer, with characteristically hydrophobic amino acids. The term "structural genomics" denotes the assignment of structure to a gene product by any means available; "functional genomics" refers to the assignment of function to a gene product.

Proteomics focuses on gene products (i.e., proteins). The primary task is to correlate the pattern of gene expression with the state of the organism. For any given (eukaryotic) cell, typically only 10 % of the genes are actually translated into proteins under a given set of conditions and at a particular epoch in the cell's life. On the other hand, a given gene sequence can give rise to tens of different proteins, by varying the arrangements of the exons (Sect. 10.7.5) and by post-translational modification. Insofar as proteins are the primary vehicle of phenotype, proteomics constitutes a communication channel between genotype and phenotype. One may think of the proteome as the "vocabulary" of the genome: Just as we use words to convey ideas and build up our individual characters, so is the genome helpless without proteins. Clearly, the proteome forms the molecular core of epigenetics. Once expression data are available, work can start on their analysis. Via the proteome, genetic regulatory networks can be elucidated.

The raw data of proteomics is either the transcriptome—a list of all the transcribed mRNAs and their abundances at a particular epoch—or the proteome—a list of all the translated proteins and their abundances, or net rates of synthesis, at a particular epoch. Given the processing that takes place between transcript and protein (Sect. 10.7.5), it is not surprising that there are often large differences between the transcriptome and proteome. Experimentally, the compiling of such a list involves separating the proteins from one another and then identifying them.

Comparison between the proteomes of diseased and healthy organisms forms the foundation of the molecular diagnosis of disease. This is just one of the many applications of bioinformatics to medicine, some of which are discussed in a chapter near the end of this Part.

Modern ecosystems management has recently acquired a strong informational content and, given its current topicality due to concerns about global warming, is discussed in a separate chapter.

An important division of proteomics deals with the interactions between proteins. It is indeed so important that a special word has been given to it—interactomics. The raw data of interactomics are a list of the affinities of each protein with every other protein in the cell, as well as nonprotein material such as lipid bilayers and polysaccharides, and DNA and RNA of course. Another division is called glycomics—the investigation of protein glycosylation.

Computational proteomics refers to the study of entire proteomes using the genome, looking, for example, for structural features such as transmembrane helices. Here the computational approach is especially important since proteins embedded in lipid membranes by three or more transmembrane helices are poorly recovered by current methods of experimental proteomics.

The investigation of protein products is called metabolomics. The metabolome comprises all of the molecules in the cell apart from proteins and DNA (lipids and polysaccharides are also usually excluded), and metabolomics is concerned with their identification, abundances and localization.

Recently, "-omics" has moved beyond the objects listed in Table 12.1 to encompass the behaviour of living organisms (ethomics), which is discussed in the chapter on phenomics. It is to be hoped that in due course the work being done to quantitatively observe behaviour and systematize the observations will meet the elegant theoretical concept of directive correlation (Sect. 9.2) in order to provide underlying, fundamental mechanisms.

The living information processor *par excellence* is of course the brain, and with the hope of bringing together the presently rather separate fields of bioinformatics and neurosciences it is the topic of a brief chapter. Information processing by individual cells such as an amoeba has recently been scrutinized and computational

methods extracted from its behaviour.[1] Conversely, biophysicochemical information processing (another rather new field) means constructing information processors using biological components such as lipid bilayer membranes and enzymes.[2]
Working integrated information processing systems are typically based on a planar
membrane of thickness L in which an enzyme is homogeneously immobilized. Their
operation can be understood through consideration of (a) transport phenomena and
(b) reaction kinetics, and coupling between them.

The general expression of flux density J for the ith molecular entity is

$$J_i = -D_i \frac{\partial c_i}{\partial x} + D_i \frac{Z_i F}{RT} c_i E \qquad (12.2)$$

where D is i's diffusivity, c its concentration, x a spatial coordinate, Z i's charge, F
the Faraday constant, and E the electric field. Mass balance is expressed by:

$$\frac{\partial c_i}{\partial t} = \frac{\partial J_i}{\partial x} + v_i \qquad (12.3)$$

where v accounts for all chemical reactions. One could write a similar expression for
charge balance but it may be assumed that in practice sufficient supporting electrolyte
is present for that to become unimportant. Combining (12.2) with (12.3) yields

$$\frac{\partial c_i}{\partial t} = D_i \frac{\partial^2 c_i}{\partial x^2} - D_i \frac{z_i F}{RT} E \frac{\partial c_i}{\partial x} + v_i \ . \qquad (12.4)$$

This is the fundamental equation of biophysicochemical information processing.[3]
Systems constructed on this basis can have fully integrated functions, unlike biosensors, in which the biosensing element is coupled to a physical transducer. Some
examples of the information processing achievable in such systems are active transport (against a concentration gradient), clocks, mathematical operations (addition,
multiplication etc.), control (e.g. stopping a function), storage and amplification (see
footnote 2).

As a more detailed example (see footnote 2), consider a membrane in which
enzyme E_1 is distributed homogeneously in the left half of a membrane separating two
compartments containing a substance S, the concentration of S being significantly
higher in the right hand compartment than in the left hand one, and enzyme E_2
distributed homogeneously in the right half of a membrane. Applying Eq. (12.4) in
the absence of an electric field ($E = 0$) yields:

$$\frac{\partial s}{\partial t} = D_S \frac{\partial^2 s}{\partial x^2} + v_2 - v_1 \ ; \qquad (12.5)$$

$$\frac{\partial p}{\partial t} = D_P \frac{\partial^2 p}{\partial x^2} + v_1 - v_2 \ . \qquad (12.6)$$

where s and p are respectively the concentrations of substrate S and product P. Some
of the solutions of this two-equation set have asymmetrical concentration profiles

---

[1] Nakagaki et al. (2009).
[2] Valleton (1990).
[3] Mostly, systems of these equations (one for each i) have to be solved numerically.

$s(x)$ and $p(x)$: for example, depletion of S in the left hand side where E_1 operates, and accumulation in the right hand side where E_2 operates. Such a profile corresponds to active transport against the concentration gradient.

References

Nakagaki T, Tero A, Kobayashi R, Onishi I, Miyaji T (2009) Computational ability of cells based on cell dynamics and adaptability. New Gener Comput 27:57–81

Valleton J-M (1990) Information processing in biomolecule-based biomimetic systems. React Polym 12:109–131

Genomics

13

We start with a couple of definitions: The genome is the ensemble of genes in an organism, and genomics is the study of the genome. The major goal of genomics is to determine the function of each gene in the genome (i.e., to annotate the sequence). This is sometimes expressly designated functional genomics. Figure 13.1 gives an outline of the topic. The starting point is the gene; we shall not deal with gene mapping, since it is already well covered in genetics textbooks. We shall view the primary experimental data of genomics as the actual nucleotide sequence and reiterate that genomics could simply be viewed as the study of the nonrandomness of DNA sequences.

The first section of this chapter will briefly review experimental DNA sequencing. The next essential step is to identify the genes. Initially, this was the sole or main preoccupation, but since then it is recognized that promoter and other sequences (including those generating small interfering RNA) possibly involved in regulation must also be considered—in brief, all biochemically active sites—since understanding of even a minimal phenotype must encompass the regulatory network controlling expression and activity, as well as the expressible genes themselves.

Once the coding sequences (i.e., the genes) have been identified, in principle one can determine the basic protein structure from the sequence alone (cf. Sect. 11.5.2). Once structure is available, function might be deduced; there is no general algorithm for doing so, but comparison with proteins of known function whose structure is already known may help to elucidate the function of new genes. It might not even be necessary to pass by the intermediate step of structure in order to deduce the function of a gene or at least to be able to make a good guess about it; merely comparing sequences of unknown function with sequences of known function, focusing on the sequence similarities, may be sufficient. The comparison of sequences of genes coding for the same (functionally speaking) protein in different species forms the basis for constructing molecular phylogenies, via their differences.

J. Ramsden, *Bioinformatics*, Computational Biology 21,
DOI 10.1007/978-1-4471-6702-0_13

Fig. 13.1 The major parts of genomics and their interrelationships. The passage from sequence to function can bypass structure via comparison with sequences of known structure

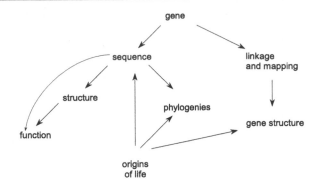

The huge collections of gene and protein data now available have encouraged the so-called "hypothesis-free" or "minimalist" approach to sequence analysis.[1] This is discussed in Sect. 13.7. Possibly the greatest value of this approach is not so much in elucidating particular phenomena such as a function of a specific gene, but rather in approaching an answer to the broader question of the meaning of the genome sequence, without the distraction of imposed categories such as "gene", which may be, as is currently all too apparent, very difficult to define unambiguously.

13.1   DNA Sequencing

The raw data used for genomic analysis are DNA sequences. This and the next section briefly describe the major experimental approaches involved. For investigating the RNA in the cell—the RNome, which has taken on a renewed importance since the discovery of the so-called "noncoding" RNA (i.e., not ultimately translated into protein), the RNA would normally first have to be converted into complementary DNA (cDNA).

---

[1]It is sometimes said of this approach, rather disparagingly perhaps, that "one can apparently make significant discoveries about a biological phenomenon without insight or intuition". Possibly this criticism derives from J.S. Mill's view that deduction cannot produce new knowledge. At any rate, it belies the fact that in reality once some unsuspected structural feature in the sequence has been discovered purely by manipulating the symbols, a great deal of insight and intuition is generally applied to make sense of it.

13.1.1 Extraction of Nucleic Acids

The following steps are typical of what is required:

1. Cell separation from the medium in which they are grown by filtration or centrifugation;
2. Cell lysis (i.e., disruption of the cell membranes, mechanically or with detergent, enzymes, etc.) and elimination of cell debris;
3. Isolation of the nucleic acids by selective adsorption followed by washing and elution.[2]

13.1.2 The Polymerase Chain Reaction

If the amount of DNA is very small, it can be multiply copied ("amplified") by the polymerase chain reaction (PCR) before further analysis. The following steps are involved:

1. Denature (separate) the two strands at 95 °C (i.e., melting).
2. Lower the temperature to 60 °C and add primer (i.e., short synthetic chains of DNA that bind at the beginning, the so-called 3′ end, of the sequence to be amplified).
3. Add DNA polymerase (usually extracted from the thermophilic microbe *Thermus aquaticus* and hence called *Taq* polymerase) and deoxyribonuclease triphosphates (dNTPs; i.e., an adequate supply of monomers); the polymerase synthesizes the complementary strand starting from the primer.
4. Stop DNA synthesis (e.g., by adding an auxiliary primer complementary to the end of the section of the template to be copied); go to step 1.

The concentration of single strands doubles on each cycle up to about 20 repetitions, after which it declines. There is of course no proofreading. Miniature bioMEMS (lab-on-a-chip) devices are now available for PCR, which operate with only a few nanolitres of solution, and enable much faster operation.

13.1.3 Sequencing

The classical technique is that devised by Sanger. One starts with many single-stranded copies of the unknown sequence, to which a known short marker sequence has been joined at one end. An oligonucleotide primer complementary to the marker is added, together with DNA polymerase and nucleotides. A small proportion of the

---

[2]This procedure may yield a preparation containing RNA as well as DNA, but RNA binds preferentially to boronate and thus can be separated from DNA.

nucleotides are fluorescently labelled dideoxynucleotides lacking the hydroxyl group necessary for chain extension. Hybridization of the primer to the marker initiates DNA polymerization templated by the unknown sequence. Whenever one of the dideoxynucleotides is incorporated, extension of that chain is terminated. After the system has been allowed to run for a time, such that all possible lengths may be presumed to have been synthesized, the DNA is separated into single strands and separated electrophoretically on a gel. The electrophoretogram (sometimes referred to as an electropherogram) shows successive peaks differing in size by one nucleotide. Since the dideoxynucleotides are labelled with a different fluorophore for each base, the successive nucleotides in the unknown sequence can be read off by observing the fluorescence of the consecutive peaks.

A useful approach for very long unknown sequences (such as whole genomes) is to randomly fragment the entire genome (e.g., using ultrasound). The fragments, approximately two megabases long and sufficient to cover the genome fivefold to tenfold, are cloned into a plasmid vector,[3] inserted into a bacterial genome and multiplied. The extracted and purified DNA fragments are then sequenced as above. The presence of overlaps allows the original sequence to be reconstructed.[4] This method is usually called shotgun sequencing. Of course, overlaps are not guaranteed, but gaps can be filled in principle by conventional sequencing.[5]

Every aspect of sequencing (reagents, procedures, separation methods, etc.) has, of course, been subject to much development and improvement since its invention (in Sanger's original method, the dideoxynucleotides were radioactively labelled), and there are now high-throughput automated methods in routine use.

Another popular technique is pyrosequencing, whereby one kind of nucleotide only is added to the polymerizing complementary chain; if it is complementary to the unknown sequence at the actual position, pyrophosphate is released upon incorporation of the complementary nucleotide. Using some other reagents, this is converted to ATP, which is then hydrolysed by the chemiluminescent enzyme luciferin, yielding a brief pulse of detectable light. The technique is suitable for automation. It is, however, practically limited to sequencing strands shorter than about 150 base pairs.

New techniques are constantly being developed, with special interest being shown in single-molecule sequencing, which would obviate the need for amplification of the unknown DNA.[6] One should also note inexpensive methods designed to detect the presence of a mutation in a sequence; steady progress in automation is enabling ever larger pieces of DNA to be tackled.

---

[3]In this context, "vector" is used in the sense of vehicle.

[4]This is somewhat related to Kruskal's multidimensional scaling (MD-SCAL or MDS) analysis.

[5]Unambiguously assembled nonoverlapping sequences are called "contigs" .

[6]See França et al. (2002) for a review, and Braslavsky et al. (2003) for a recent single-molecule technique.

13.1.4 Expressed Sequence Tags

Expressed sequence tags (ESTs) are derived from the cDNA complementary to mRNA. They consist of the sequence of typically 200–600 bases of a gene, sufficient to uniquely identify the gene. The importance of ESTs is, however, tending to diminish as sequencing methods become more powerful.

Expressed sequence tags are generated by isolating the mRNA from a particular cell line or tissue and reverse-transcribing it into cDNA, which is then cloned into a vector to make a "library".[7] Some 400 bases from the ends of individual clones are then sequenced.

If they overlap, ESTs can be used to reconstruct the whole sequence as in shotgun sequencing, but their primary use is to facilitate the rapid identification of DNA. For various reasons, not least low-fidelity transcription, the sequences are typically considerably less reliable than those generated by conventional gene sequencing.

13.2 DNA Methylation Profiling

Although the overall proportion of methylated DNA can be determined chemically, in order to properly understand the regulatory rôle of methylation, it is necessary to determine the methylation status of each base in sequence (bearing in mind that only CpG is methylated). The methylation status of a nucleotide can be determined by pyrosequencing (Sect. 13.1.3), but the technique is limited to relatively short nucleotide sequences. A more recent method relies on treating DNA with bisulfite (under acidic conditions cytosine is converted to uracil, and methylated cytosine is not) and comparing the sequence with the untreated one.[8] Even newer is the technique called MethylCap-seq:[9] The DNA is sonicated, fragmenting it to pieces with a length of around 300 base pairs, which are then exposed to MBD-GST immobilized on magnetic beads, which captures methylated fragments at low concentrations of NaCl; a gradient of increasing salt concentration elutes the DNA fragments from the beads. Epigenetic profiling is of growing importance to medicine.[10]

13.3 Gene Identification

The ultimate goal of gene identification (or "gene prediction") is automatic annotation: to identify all biochemically active portions of the genome by algorithmically processing the sequence and to predict the reactions and reaction products of those

---

[7]In this context, "library" is used merely to denote "collection".

[8]Bibikova et al. (2006); Bibikova and Fan (2010).

[9]Brinkman et al. (2010); for other methods see Zuo et al. (2009).

[10]See, e.g., Heyn and Esteller (2012).

portions coding for proteins. At present we are still some way from this goal. Success will not only allow one to discover the functions of natural genes but should also enable the biochemistry of new, artificial sequences to be predicted and, ultimately, to prescribe the sequence necessary to accomplish a given function.

In eukaryotes, the complicated exon–intron structure of the genome makes it particularly difficult to predict the course of the key operations of transcription, splicing, and translation from sequence alone (even without the possibility that essential instructions encoded in acylation of histones, etc. are transmitted epigenetically from generation to generation).

Challenges remain in identifying the exons, introns, promoters, and so on in each stretch of DNA, such that the exons could be grouped into genes and the promoters assigned to the genes or groups of genes whose transcription they control. Other tasks include the identification of those genes (in humans, mammals, etc.) believed to originate from viruses and the localization of hypervariable regions (e.g., those coding for immunoglobulins). Ultimately, the aim is to be able to understand the relationships among the various elements of the genome.

Gene prediction can be divided into intrinsic (template) and extrinsic (lookup) methods. The former are the best candidates for leading to fundamental insight into how the gene works; if they are successful, they should furthermore then inevitably provide the means to generalize from the biochemistry of natural sequences to yield rules for designing new genes (and genomes) to fulfil specified functions. We shall begin, however, by considering the conceptually simpler extrinsic methods.

13.4   Extrinsic Methods

The principle of the extrinsic or lookup method is to identify a gene by finding a sufficiently similar known object in existing databases. Hence, the method is based on sequence similarity (to be discussed in Sect. 13.4.2), using the still relatively small core of genes identified by classical genetic and molecular biological studies to prime the comparison; that is, a gene of unknown function is compared with the database of sequences with known function. This approach reflects a widely used, but not necessarily correct (or genuinely useful), assumption that similar sequences have similar functionality.[11] A major limitation of this approach is the fact that, at present, about a third of the sequences of newly sequenced organisms turn out to match no sufficiently similar known sequences in existing databanks. Furthermore, errors in the sequences deposited in databases can pose a serious problem.

---

[11]Note that "homology" is defined as "similarity in structure of an organ or molecule, reflecting a common evolutionary origin". Sequence similarity is insufficient to establish homology, since genomes contain both orthologous (related via common descent) and paralogous (resulting from duplications within the genome) genes.

13.4.1 Database Reliability

An inference, especially a deductive one, drawn from data is only as good as the data from which it is formed. The question of the reliability of the data is certainly a matter for legitimate concern. The most pernicious errors are wrong nucleic acid bases in a sequence. The sources of such errors are legion and range from experimental uncertainties to mistakes in typing the letters into a file using a keyboard. Of course, these errors can be considered as a source of noise (i.e., equivocation) and handled with the ideas developed earlier, especially in Chap. 3. Undoubtedly there is a certain redundancy in the sequences, but these questions of equivocation and redundancy in database sequences and the consequences for deductive inference do not yet seem to have been given the attention they deserve. In particular, there appears to be a feeling associated with the "big data" movement that, provided one has enough data, the errors will somehow be "averaged out" or "autocompensated", although proper justification for this notion is lacking.

13.4.2 Sequence Comparison and Alignment

The pairwise comparison of sequences is very widely used in bioinformatics. Evidently, it is a subset of the general problem of pattern recognition (Sect. 8.2). If it were only a question of finding matches to more or less lengthy blocks of symbol sequences (e.g., the longest common subsequence; LCS), the task would be relatively straightforward and the main work would be merely to assess the statistical significance of the result; that is, compare with the null hypothesis that a match occurred by chance (cf. Sect. 5.2.1). In reality, however, the two sequences one is trying to compare differ due to mutations, insertions, and deletions (cf. Sect. 10.6.1), which renders the problem considerably more complicated; one has to allow for gaps, and one tries to make inferences from local alignments between subsequences. A typical example of an attempt to align fragments of two nucleotide sequences is

$$
\begin{array}{c}
\mathrm{A\,C\,G\,T\,A\,C\,G\,T\,A - G\,T} \\
|\ |\quad\ \ |\ \ |\ |\ |\ \ |\ | \\
\mathrm{A\,C - - A\,T\,G\,T\,A\,C\,G\,T}
\end{array}
$$

where vertical lines indicate matches. Note the gaps that have been inserted to increase the number of matches. In the absence of gaps, one could simply compute the Hamming distance between two sequences; the introduction of the possibility of gaps introduces two problems: (i) the number of possible alignments becomes very large and (ii) where are gaps to be placed in sequence space?

If no gaps are allowed, one assigns and sums scores for all possible pairs of aligned substrings within the two sequences to be matched. If gaps are allowed, there are $\binom{2n}{n}$

possible alignments of two sequences each of length n.[12] Even for moderate values of n there are too many possibilities to be enumerated (problem (i), a computational one). It is solved using dynamic programming algorithms (Sect. 13.4.4). Problem (ii) is solved by devising a scoring system with which gaps and substitutions can be assigned numerical values. Finally, one needs to assess the statistical significance of the alignment. This is still an unsolved problem—let us call it problem (iii).

The essence of sequence alignment is to assign a score, or cost, for each possible alignment; the one with the lowest cost, or highest score, is the best one, and if aligning multiple sequences, degrees of kinship can be assigned on the basis of the score, which has the form

$$\text{total score} = \text{score for aligned pairs} + \text{score for gaps} . \qquad (13.1)$$

The score is, in effect, the relative likelihood that a pair of sequences are related. It represents distance, together with the operations (mutations and introduction of gaps) required to edit one sequence onto the other. Sequence alignment attempts to maximize the number of matches while minimizing the number of mutations and gaps required in the editing process. Unfortunately, the relative weights of the terms on the right-hand side of (13.1) are arbitrary. The main approach to assigning weights to the terms more objectively is to study many extant sequences from organisms one knows from independent evidence to be related. In principle, under a given set of conditions (e.g., a certain level of exposure to cosmic rays), a given mutation presumably has a definite probability of occurrence; that is, it can, at least in principle, be derived from an objective set of data according to the frequentist interpretation, but the practical difficulties and the possibility that such probabilities may be specific to the sequence neighbouring the mutation make this an unpromising approach.

Whereas with DNA sequences, a nucleotide is—at least to a first approximation— either matched or not, with polypeptides a substitution might be sufficiently close chemically so as to be functionally neutral. Hence, if alignments are carried out at the level of amino acids, exact matches and substitutions are dealt with by compiling an empirical table, based on chemical or biological knowledge or both, of degrees of equivalence.[13] There is no uniquely optimal table. To construct one, a good starting point is the table of amino acids (Table 11.6). Isoleucine should have about the same score for substitution by leucine as for an exact match and so forth; substitution of a polar for an apolar group or lysine for glutamic acid (say) would be given low or negative scores. The biological approach is to look at the frequencies of the different substitutions in pairs of proteins that can be considered to be functionally equivalent from independent evidence (e.g., two enzymes that catalyse the same reaction).

In essence, the entries in a scoring matrix are numbers related to the probability of a residue occurring in an alignment. Typically, they are calculated as (the logarithm

---

[12]This is obtained by considering the number of ways of intercalating two sequences while preserving the order of symbols in each.

[13]For example, BLOSUM50, a 20×20 score matrix (histidine scores 10 if replacing histidine, glutamine 0, alanine –3, and so on). The diagonal terms are not equal.

of) the probability of the "meaningful" occurrence of a pair of residues divided by the probability of random occurrence. Probabilities of "meaningful" occurrences are derived from actual alignments "known to be valid". The inherent circularity of this procedure gives it a temporary and provisional air.

In the case of gaps, the (negative) score might be a single value per gap or could have two parameters: one for starting a gap, and another, multiplied by the gap length, for continuing it (called an affine gap cost). This takes some slight account of possible correlations in the history of changes presumed to have been responsible for causing the divergence in sequences. The scoring of substitutions considers each mutation to be an independent event, however.

In summary, the central themes of sequence comparison are:[14] distance functions appropriate in the absence of natural correspondence of elements; optimum correspondences between sequences; and dynamic programming algorithms (Sect. 13.4.4) for calculating the distances and optimum correspondences.

13.4.3 Trace, Alignment and Listing

These are, perhaps, the three most important modes of presentation for the analysis of differences between sequences. Trace consists of the source sequence above and the target sequence below, with lines, at most one per element and not crossing each other, from some elements in the source to some in the target. The lines provide at least a partial correspondence between source and target. There are two kinds of matches of a pair: if the connected elements are the same, they are referred to as an identity or a continuation; if they are different, a substitution. A source element without a line is referred to as a deletion; a target element, an insertion (the term *indel* means either an insertion or a deletion). This is illustrated below.

I N D U S T R Y

I N T E R E S T
Note the substitution of Y by S.

Problem. Construct as many different analyses as possible of the above pair of sequences using trace.

An alignment or matching consists of, again, the source sequence above and the target below, forming a two-row matrix. Both rows can be interspersed with null characters (represented by Ø—or simply a blank)—note that a column of null characters is not permitted. Deletion has the null character below; a column with the null character above is a substitution. The absence of Ø denotes a match; if the

---

[14] Kruskal (1964), Chap. 1 of Sankoff and Kruskal (1999).

elements are equal it is a continuation, if unequal a substitution:

$$\begin{bmatrix} I\ N\ D\ U\ S\ T\ \emptyset\ R\ \emptyset\ Y\ \emptyset \\ I\ N\ \emptyset\ \emptyset\ \emptyset\ T\ E\ R\ E\ S\ T \end{bmatrix}$$

Problem. Construct as many different analyses as possible of the above pair of sequences using alignment.

A third mode is called a listing or derivation; it consists of an alternating series of sequences and elementary operations, successive sequences differing only in accord with the interspersed elementary operation, as illustrated below:

INDUSTRY

delete D

INUSTRY

delete U

INSTRY

substitute Y by S

INSTRS

insert E

INSTERS

insert E

INSTERES

delete S

INTERES

insert T

INTEREST

Listing is of less practical use, but is a richer mode of analysis than the previous two.

13.4.4 Dynamic Programming Algorithms

The concept of dynamic programming comes from operations research, where it is commonly used to solve problems that can be divided into stages with a decision required at each stage. A good generic example is the problem of finding the shortest path on a graph. The decisions are where to go next at each node. It is characteristic that the decision at one stage transforms that state into a state in the next stage. Once that is done, from the viewpoint of the current state the optimal decision for the remaining states does not depend on the previous states or decisions. Hence, it is not necessary to know how a node was reached, only that it was reached. A recursive relationship identifies the optimal decision for stage M, given that stage $M + 1$ has already been solved; the final stage must be solvable by itself.

The following is a generic dynamic programming algorithm (DPA) for comparing two strings $S1$ and $S2$ with $M[i, j] = $ cost or score of $S1[1..i]$ and $S2[1..j]$:[15]

```
M[0, 0] = z
for each i in 1 .. S1.length
  M[i,0] = f( M[i-1, 0 ], c(S1[i],"_" ) )         -- Boundary

for each j in 1 .. S2.length
  M[0,j] = f( M[0,  j-1], c("_", S2[j] ) )         -- conditions

for each i in 1 .. S1.length and j in 1 .. S2.length
  M[i,j] = g(f(M[i-1, j-1], c(S1[i], S2[j])),   -- (mis)match
             f(M[i-1, j ], c(S1[i], "_" )),     -- delete S1[i]
             f(M[i,  j-1], c("_",   S2[j])))    -- insert S2[j]
```

Applied to sequence alignment, two varieties of DPA are in use: the Needleman–Wunsch ("global alignment") algorithm which builds up an alignment starting with easily achievable alignments of small subsequences, and the Smith–Waterman ("local alignment") algorithm that is similar in concept, except that it does not systematically move through the sequences from one end to the other, but compares subsequences anywhere.

It is often tacitly assumed that the sequences are random (i.e., incompressible), but if they are not (i.e., they are compressible to some degree), this should be taken into account.

There are also some heuristic algorithms (e.g., BLAST and FASTA) that are faster than the DPAs. They look for matches of short subsequences, which may be only a few nucleotides or amino acids long, that they then seek to extend. As with the DPAs, some kind of scoring system has to be used to quantify matches.

Although sequence alignment has become very popular, some of the assumptions are quite weak and there is strong motivation to seek alternative methods for evaluating the degree of kinship between sequences, not based on symbol-by-symbol comparison; for example, one could evaluate the mutual information between strings a and b:

$$I(s_a, s_b) = I(s_b, s_a) = I(s_a) - I(s_a|s_b) = I(s_b) - I(s_b|s_a) . \tag{13.2}$$

Multiple alignment is an obvious extension of pairwise alignment.

13.5   Intrinsic Methods

The template or intrinsic approach involves constructing concise descriptions of prototype objects and then identifying genes by searching for matches to such prototypes. An elementary example is searching for motifs (i.e., short subsequences)

---

[15]Allison et al. (1999).

known to interact with particular drugs. The *motif* is often defined more formally along the lines of a sequence of amino acids that itself defines a substructure in a protein that can be connected in some way to protein function or structural stability and, hence, that appears as conserved regions in a group of evolutionarily related gene sequences. This is not a strong definition, not least because the motif concept is based on a mosaic view of the genome that is opposed to the more realistic (but less tractable) systems view.

The construction of the concise descriptions could be either deductive or inductive. A difficulty is that extant natural genomes are not elegantly designed from scratch, but assembled *ad hoc*, and refined by "life experience" (of the species). The use of fuzzy criteria may help to overcome this problem.

In practice, intrinsic methods often boil down to either computing one or more parameters from the sequence and comparing them with the same parameters computed for sequences of known function, or searching for short sequences that experience has shown are characteristic of certain functions.

13.5.1 Signals

In the context of intrinsic methods for assigning function to DNA, the term "signal" denotes a short sequence relevant to the interaction of the gene expression machinery with the DNA. In effect, one is paralleling the action of the cell (e.g., the transcription, splicing, and translation operations) by trying to recognize where the gene expression machinery interacts with DNA. In a sense, therefore, this topic belongs equally well to interactomics (Chap. 16). Much use has been made of so-called consensus sequences, which are formed from sequences well conserved over many species by taking the most common base at each position. The distance (e.g., the Hamming distance) of an unknown sequence from the consensus sequence is then computed; the closer they are, the more likely it is that the unknown sequence has the same function as that represented by the consensus sequence. Useful signals include start and stop codons (Table 3.1). More sophisticated signals include sequences predicted to result in unusual DNA bendability or known to be involved in positioning DNA around histones, intron splice sites in eukaryotic pre-mRNA and sequences corresponding to ribosome binding sites on RNA, and so on.

Special effort has been devoted to identifying promoters, which are of great interest as potential targets for new drugs. It is a hard problem because of the large and variable distances between the promoter(s) and the sequence to be transcribed. The approach relies on relatively well conserved sequences (i.e., effectively consensus sequences) such as TATA or CCAAT. Other sites for protein–DNA interactions can be examined in the same way; indeed, the entire transcription factor binding site can be included in the prototype object, which allows more sophistication (e.g., some constraints between the sequences of the different parts) to be applied.

13.5.2 Hidden Markov Models

Knowledge of the actual biological sequence of processing operations can be used to exploit the effect of the constraints on (nucleic acid) sequence that these successive processes imply. One presumes that the Markov binary symbol transition matrices are slightly different for introns, exons, promoters, enhancers, the complementary strand, and so forth. One constructs a more elaborate automaton, an automaton of automata, in which the outer one controls the transitions between the different types of DNA (introns, exons, etc.) and the inner set gives, for each type, the 16 different binary transition probabilities for the symbol sequence. More sophisticated models use higher-order chains for the symbol transitions; further levels of automata can also be introduced. The epithet "hidden" is intended to signify that only transitions from symbol to symbol are observable, not transitions from type to type. The main problem is the statistical inadequacy of the predictions. A promoter may only have two dozen bases; a fourth-order Markov chain for nucleotides has of the order of 10^{10} transition probabilities.

Problem. Construct a hidden Markov model for the mitogen-activated protein kinase signalling cascade (Sect. 14.7).

13.6 Beyond Sequence

Proteomics data (see Chaps. 14 and 16) are integrated with sequence information in the attempt to assign function. Proteins whose mRNA levels are correlated with each other, proteins whose homologues are fused into a single gene in some organisms, those which have evolved in a correlated fashion, those whose homologues operate together in a metabolic path or that are known to physically interact can all be considered to be linked in some way; for example, a protein of unknown function whose expression profile (see footnote 6 in Chap. 14) matches that of a protein of known function in another organism is assigned the same function. In a literary analogy, one could rank the frequencies of words in an unknown and known language and assign the same meanings to the same ranks. Whether the syntax of gene expression is sufficiently shared by all organisms to allow this to be done reliably is presently an open question.

Other kinds of data assisting protein function prediction are structure prediction, intracellular localization, signal peptide cleavage sites of secreted proteins, glycosylation sites, lipidation sites, phosphorylation sites, other sites for post-translational modification, cofactor binding sites, dehydron density, and so on.

13.7   Minimalist Approaches

The inspiration for this approach is the study of texts written in human languages. A powerful motivation for the development of linguistics as a formal field of inquiry was the desire to understand texts written in "lost" languages (without living speakers), especially those of antiquity, records of which began pouring into Europe as a result of the large-scale expeditions to Egypt, Mesopotamia, and elsewhere undertaken in the nineteenth and twentieth centuries. More recently, linguistics has been driven by attempts to automatically translate texts written in one language into another.

One of the most obvious differences between DNA sequences and texts written in living languages is that the former lacks separators between the words (denoted by spaces in most of the latter). Furthermore, unambiguous punctuation marks generally enable phrases and sentences in living languages to be clearly identified. Even with this invaluable information, however, matters are far from determined and the study of the morphology of words and the rules that determine their association into sentences (syntax)—that is, grammar—is a large and active research field.

For DNA that is ultimately translated into protein sequences, the nucleic acid base pairs are grouped into triplets constituting the reading frames, each triplet corresponding to one amino acid. A further peculiarity of DNA compared with human languages is that reading frames may overlap; that is, from the sequence AAGTTCTG... one may derive the triplets AAG, AGT, GTT, TTC, ... This is encountered in certain viruses,[16] which generally have very compact genomes. However, the reading frames of eukaryotes are generally nonoverlapping (i.e., only the triplets AAG, TTC, ... would be available).

Due to the absence of unambiguous separators, the available structural information in DNA is much more basic than in a human language. Even if the "meaning" of a DNA sequence (a gene) that corresponds to a functional protein might be more or less clear, especially in the case of prokaryotes, it must be remembered that the sequence may be shot through with introns; even the stop codons (Table 3.1) are not unambiguous. Only a small fraction (a few percent) of eukaryotic genome sequences actually correspond to proteins, and any serious attempt to understand the semantics of the genome must encompass the totality of its sequence.

Nucleotide Frequencies

Due to the lack of separators, it is necessary to work with n-grams rather than words as such. Basic information about the sequence is encapsulated in the frequency dictionaries W_n of the n-grams, (i.e., lists of the numbers of occurrences of each possible n-gram). Each sequence can then be plotted as a point in M^n-dimensional space, where M is the number of letters in the alphabet (=4 for DNA, or 5 if we include methylated cytosine as a distinct base).

---

[16]E.g., Zaaijer et al. (2007).

Even such very basic information can be used to distinguish between different genomes; for example, thermophilic organisms are generally richer in C and G, because the C–G base-pairing is stronger and, hence, stabler at higher temperatures than A–T. Furthermore, since each genome corresponds to a point in a particular space, distances between them can be determined, and phylogenetic trees can be assembled.

The four-dimensional space corresponding to the single base-pair frequencies is not perhaps very interesting. Already the 16-dimensional space corresponding to the dinucleotide frequencies is richer and might be expected to be more revealing. In particular, given the single base-pair frequencies, one can compute the dinucleotide frequencies expected from random assembly of the genome and determine divergences from randomness. Dinucleotide bias is assessed, for example, by the odds ratio $\rho_{XY} = w_{XY}/(w_X w_Y)$, where w_X is the frequency of nucleotide X.[17] We will return to this comparison of actual with expected frequencies below.

Instead of representing the entire genome by a single point, one can divide it up into roughly gene-long fragments (100–1000 base pairs), determine their frequency dictionaries, and apply some kind of clustering algorithm to the collection of points thereby generated. Alternatively, dimensional reduction using principal component analysis (Sect. 8.3.2) may be adequate. The distributions of a single base-pair and dinucleotide frequencies look like Gaussian clouds, but the triplet frequencies reveal a remarkable seven-cluster structure.[18] It is natural to interpret the seven clusters as the six possible reading frames (three in each direction) plus the "noncoding" DNA.

Word Occurrences

Once the single-nucleotide frequencies are known, it is possible to calculate the expectations of the frequencies of n-grams assembled by random juxtaposition. Constraints on the assembly are revealed by deviations of the actual frequencies from the expected values. This is the principle of the determination of dinucleotide bias. It is, however, limited with regard to the inferences that may be drawn. For one thing, as n increases, the statistics become very poor. The genome of *E. coli*, for example, is barely large enough to contain a single example of every possible 11-gram even if each one was deliberately included. Furthermore, the comparison of actual frequencies with expected ones depends on the model used to calculate the expected frequencies. All higher-order correlations are subsumed into a single number, from which little can be said about the relative importance of a particular sequence.

It is possible to approach this problem more objectively (according to a maximum entropy principle[19]) by asking what is the most probable continuation of a given n-

---

[17]See, e.g., Karlin et al. (1994).
[18]Gorban et al. (2005).
[19]The entropy of a frequency dictionary is defined as

$$S_n = -\sum_{j=1} f_j \log f_j.$$

(13.3)

gram (cf. Eq. 2.21). Frequency dictionaries may be reconstructed from thinner ones according to this principle; for example, if one wishes to reconstruct the dictionary W_n from W_{n-1}, the reconstructed frequencies are[20]

$$\tilde{f}_{i_1,\dots,i_n} = \frac{f_{i_1,\dots,i_{n-1}} f_{i_2,\dots,i_n}}{f_{i_2,\dots,i_{n-1}}} , \tag{13.4}$$

where $i_1, \dots$ are the successive nucleotides in the n-gram. The reconstructed dictionary is denoted by $\tilde{W}_n(n-1)$. The most unexpected, and hence informative, n-grams are then those with the biggest differences between the real and reconstructed frequencies (i.e., with values of the ratio $f/\tilde{f}$ significantly different from unity).

13.8 Phylogenies

The notion that life-forms evolved from a single common ancestor (i.e., that the history of life is a tree) is pervasive in biology.[21] Before gene and protein sequences became available, trees were constructed from the externally observable characteristics of organisms. Each organism is therefore represented by a point in phenotype space. In the simplest (binary) realization, a characteristic is either absent (0) or present (1) or is present in either a primitive (0) or an evolved (1) form. The distance between species, compared in pairs, can be computed as a Hamming distance (i.e., the number of different characteristics); for example, consider three species A, B, and C, to which 10 characteristics labelled a to j are assigned:

$$
\begin{array}{c|cccccccccc}
 & a & b & c & d & e & f & g & h & i & j \\
\hline
A & 1 & 1 & 1 & 1 & 1 & 1 & 1 & 0 & 0 & 1 \\
B & 0 & 0 & 0 & 0 & 0 & 1 & 1 & 1 & 0 & 0 \\
C & 0 & 0 & 0 & 0 & 0 & 0 & 0 & 0 & 1 & 0
\end{array} \tag{13.5}
$$

This yields the symmetric distance matrix

$$
\begin{array}{c|ccc}
 & A & B & C \\
\hline
A & 0.0 & & \\
B & 0.7 & 0.0 & \\
C & 0.9 & 0.4 & 0.0
\end{array} \tag{13.6}
$$

The species are then clustered; the first cluster is formed from the closest pair (viz. B and C in this example) and the next cluster is formed between this pair and the species

---

[20]Gorban et al. (2000).
[21]The concept of phylogeny was introduced by E. Haeckel; see Sect. 10.8.

closest to its two members (and so forth in a larger group) to yield the following tree or dendrogram:

$$- - \Big|^{\begin{array}{c} - - - - \\ \end{array}} \Big|^{\begin{array}{c} - - - - \ B \\ - - - - \ C \end{array}} .$$

$$- - - - - - - - \ A \tag{13.7}$$

This is the classical method; the root of the tree is the common ancestor.

An alternative method, called cladistics,[22] counts the number of transformations necessary to go from a primitive to an evolved form. Hence, in the example, C differs by just one transformation from the putative primitive form (all zeros). Two transformations (of characters f and g) create a common ancestor to A and B, but it must be on a different branch from that of C, which does not have evolved forms of those two characteristics. This approach yields a different tree:

$$- - \Big|^{\begin{array}{c} - - - - - \ A \\ - \ B \end{array}}$$

$$- \ C \tag{13.8}$$

The principle of construction of a *molecular phylogeny* is to use the sequences of the "same" genes (i.e., encoding a protein of the same function) in different organisms as the characteristic of the species; that is, molecular phylogenies are based on genotype rather than phenotype. In actual practice, protein sequences are typically used, which are intermediate between genotype and phenotype. In the earliest studies (1965–1975), cytochrome c was a popular object, since it is found in nearly all organisms, from bacteria to man. Later, the sequence of the small subunit of ribosomal RNA (rRNA), another essential and universal object, was used.[23] Nowadays, one can, in principle, analyse whole genomes.

A chronology can be established on the premiss that the more changes there are, the longer the time elapsed since the species diverged (assuming that the changes occur at a constant rate with respect to sidereal time). This premiss can be criticized since, although the unit of change is the nucleotide, selection (the engine of speciation) acts on the amino acid; some nucleotide mutations lead to no change in amino acid due to the degeneracy of the code. There is actually little real evidence that mutations occur at random (i.e., with respect to both the site and the type of mutation).

A difficulty with molecular phylogenies is the fact that lateral gene transfer (LGT; cf. Sect. 10.6.4), especially between bacteria and between archaea, may vitiate the calculated distances. A plausible counterargument in favour of the use of rRNA is that it should be unaffected by LGT, due to its fundamental place in cell metabolism.

A further difficulty is a computational one: that of finding the optimal tree, since usually one is interested in comparing dozens (and ultimately millions) of species.

---

[22] A *clade* is a taxonomic group comprising a single common ancestor and all its descendants (i.e., a monophyletic group). A clade minus subclade(s) is called a paraphyletic group.
[23] rRNA has been championed by C. Woese.

The basic principle applied to address this problem is that of parsimony: One seeks to construct the tree with the least possible number of evolutionary steps. Unfortunately, this is an NP-complete problem and hence the computation time grows exponentially with the number of species; even a mere 20 species demands the analysis of almost 10^{22} possible trees!

13.9 Metagenomics

The ability to culture bacteria in the laboratory, of which Pasteur seems to have been the pioneer, was a crucial step in the emergence of bacteriology. Pasteur used liquid media (broths); for the purposes of investigation, however, solid media, introduced by Koch, are more convenient. Since then, the culture of bacteria has become indispensable to vast areas of medicine, biotechnology and research; for the identification and counting of bacteria and for the development of serological assays and vaccines, to name just a few ways of making use of bacterial cultures. At the same time it is recognized that the vast majority of bacteria cannot be cultured. Therefore, the only way that this extant, almost immanent microbial richness can be accessed is by sequencing its genomic signature. The genomes of the entire natural microbiota collectively constitute the metagenome.[24]

The vast increase in DNA sequencing capability—both in terms of hardware and in algorithms for analysing the raw data—has allowed metagenomics (the study of the metagenome) to become a practical science. Work begins with the extraction of the DNA of all the microbes in some environmental sample (e.g., soil, or seawater, or indeed the human gastrointestinal tract, cf. Sect. 15.3). It may then be necessary to clone the DNA (using cultured, laboratory bacteria!) in order to produce sufficient material for further analysis.

The most basic analysis is simply to sequence all the DNA. This may result in millions of genes, which can be compared with known sequences; early work appeared to reveal an astonishing diversity of whole classes of hitherto unknown genes. One should, however, be mindful of the influence of sequencing errors in giving the appearance of more novelty than is actually the case.[25]

Function-based metagenomics obviates the need to sequence the DNA by letting the fragments be translated, again in laboratory-cultured bacteria. Novel proteins of phenotypes are then further analysed.

---

[24]Rondon et al. (2000).
[25]Quince et al. (2009).

References

Allison L, Powell D, Dix TI (1999) Compression and approximate matching. Comput J 42:1–10

Bibikova M, Fan JB (2010) Genome-wide DNA methylation profiling. WIREs Syst Biol Med 2:210–223

Bibikova M, Lin Zh, Zhou L et al (2006) High-throughput DNA methylation profiling using universal bead arrays. Genome Res 16:383–393

Braslavsky I, Hebert B, Kartalov E, Quake SR (2003) Sequence information can be obtained from single DNA molecules. Proc Natl Acad Sci USA 100:3960–3964

Brinkman AB, Simmer F, Ma K, Kaan A, Zhu J, Stunnenberg HG (2010) Whole-genome DNA methylation profiling using MethylCap-seq. Methods, vol 52, pp 232–236

França LTC, Carrilho E, Kist TB (2002) A review of DNA sequencing techniques. Q Rev Biophys 35:169–200

Gorban AN, Popova TG, Sadovsky MG (2000) Classification of symbol sequences over their frequency dictionaries. Open Syst Inf Dynam 7:1–17

Gorban AN, Popova TG, Zinovyev A (2005) Codon usage trajectories and 7-cluster structure of 143 complete bacterial genomic sequences. Physica A 353:365–387

Heyn H, Esteller M (2012) DNA methylation profiling in the clinic: applications and challenges. Nature Rev Genetics 13:679–692

Karlin S, Ladunga I, Blaisdell BE (1994) Heterogeneity of genomes: measures and values. Proc Natl Acad Sci USA 91:12837–12841

Kruskal JB (1964) Multidimensional scaling by optimizing goodness of fit to a nonmetric hypothesis. Psychometrika 29:1–27

Quince C et al (2009) Accurate determination of microbial diversity from 454 pyrosequencing data. Nat Methods 6:639–641

Rondon MR et al (2000) Cloning the soil metagenome: a strategy for accessing the genetic and functional diversity of uncultured microorganisms. Appl Environ Microbiol 66:2541–2547

Sankoff D, Kruskal J (1999) Time warps, string edits, and macromolecules. CSLI Publications

Zaaijer HL, van Hemert FJ, Koppelman MH, Lukashov VV (2007) Independent evolution of overlapping polymerase and surface protein genes of hepatitis B virus. J Gen Virol 88:2137–2143

Zuo T, Tycko B, Liu T-M, Lin H-JL, Huang TH-M (2009) Methods in DNA methylation profiling. Epigenomics 1:331–345

Proteomics

14

The proteome is the ensemble of expressed proteins in a cell, and proteomics is the study of that ensemble (i.e., the identification and determination of the amounts, locations and interactions of all the proteins). The tasks of proteomics are summarized in Fig. 14.1.

We have seen in Chap. 10 how the gene is first transcribed into messenger RNA (mRNA), and a given gene, especially in a eukaryotic cell in which the gene resembles a mosaic of introns (I) and exons (E), can be assembled to form different mRNAs (e.g., if the gene is $E_1 I E_2 I E_3 I E_4 I E_5$, one could form mRNAs $E_1 E_2 E_3 E_4 E_5$, $E_1 E_3 E_4 E_5$, $E_1 E_3 E_5$, etc.). The ensemble of these transcripts is called the transcriptome, and its study is called transcriptomics (Sect. 14.1). Due to the variety of assembly possibilities, the transcriptome is considerably larger (i.e., contains more types of objects) than the genome.

After the mRNA is translated into a protein, the polypeptide may be modified by the following:

1. Cutting off a block of amino acids from either end;
2. Covalently adding a large chemical group to an amino acid (e.g., a fatty acid or an oligosaccharide);
3. Covalently modifying an amino acid (e.g., by serine or threonine phosphorylation, or acetylation);
4. Oxidizing or reducing an amino acid (e.g., arginine deimination or glutamine deamidation).

Modifications 2 and 3 may well be reversible; that is, there may be a pool of both modified and unmodified forms in the cell at any instant. More than 200 post-translational modifications (PTM) have been identified. They can significantly change the conformation, which in turn implies (for example) that the catalytic activity of an enzyme may change. Conformation is also a determinant of intermolecular specificity, which in turn determines protein binding, localization, and so forth. These are all crucial aspects of the dynamical system that precedes the phenotype in the overall genotype $\rightarrow$ phenotype transformation.

© Springer-Verlag London 2015
J. Ramsden, *Bioinformatics*, Computational Biology 21,
DOI 10.1007/978-1-4471-6702-0_14

Fig. 14.1 The major parts of proteomics and their interrelationships, and their relation to pharmaceutical drugs

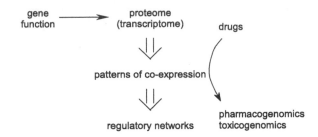

These modifications increase the potential repertoire of proteins expressible from genes by typically one to two orders of magnitude (since many combinations are possible) compared with the repertoire of genes. Notice that effecting these modifications requires enzymes; hence, the proteome is highly self-referential.

Although the number of different proteins therefore far exceeds the number of genes, the actual number of proteins present in a cell at any one instant may well be much smaller than the number of genes, since only a part of the possible repertoire is likely to be expressed. Each cell type in an organism has a markedly different proteome. The proteome for a given cell type is, moreover, likely to depend on its environment; unlike the genome, therefore, which is relatively static, the proteome can be highly dynamic.

Proteomics is sometimes defined so as to encompass what is otherwise called interactomics: the study of the ensemble of molecular interactions, especially protein–protein interactions, in a cell, including those that lead to the formation of more or less long-lived multiprotein complexes. These aspects are covered in Chap. 16.

14.1 Transcriptomics

The goal of transcriptomics is to identify, quantify, and analyse the amounts of all the mRNA in a cell. This is mainly done using microarrays ("gene chips"). The principle of a microarray is to coat a flat surface with spots of DNA complementary to the expressed mRNA, which is then captured because of the complementary base-pairing (hybridization) between DNA and RNA (A–U, C–G, G–C, T–A) and identified. The relationship of a microarray to a classical affinity assay resembles that of a massively parallel processor to a classical linear processor, in which instructions are executed sequentially. The parent classical assay is the Northern blot.[1] Microarrays consist of

---

[1] Northern blotting allows detection of specific RNA sequences. RNA is fractionated by agarose gel electrophoresis, followed by transfer (blotting) to a membrane support, followed by hybridization with known DNA or RNA probes that are radioactively or fluorescently labelled to facilitate their detection. The technique can be thought of as a variant of Southern blotting, in which specific DNA sequences from a sample are probed in a similar fashion.

Table 14.1 Typical features of microarrays

Application	Capture element	Sample
Genomics	ESTs	DNA
Transcriptomics	cDNA	mRNA
Proteomics	Antibodies	Proteins
Metabolomics	Various	Various

a two-dimensional array, typically a few square millimetres in overall area, of more or less contiguous patches, the area of each patch being a few square micrometres (or less) and each patch on the array having a different chemical composition. Typical microarrays are assembled from one type of substance (e.g., nucleic acid oligomers).

In use, the array is flooded with the sample whose composition one is trying to elucidate.[2] After some time has elapsed, the array is scanned to determine which patches have captured something from the sample. It is, of course, essential that each patch should be addressable, in the sense that the composition of each individual patch is known or traceable. Hence, a photomicrograph of the array after exposure to the analyte should allow one to determine which substances have been captured from the sample.

Table 14.1 summarizes some features of microarrays. In more detail, the protocol for a microarray assay would typically involve the following steps:

Array preparation. The chip should be designed on the basis of what one is looking for. Each gene of interest should be represented by at least one, or preferably more, unique subsequences.[3] Once the set of sequences has been selected, there are two main approaches to transfer them to the chip:

1. Heterooligomers complementary to the mRNA of interest are assembled from successive monomers using microfabrication technology; for example,[4] photoactivatable nucleic acid monomers are prepared. Exposure through a mask, or with a laser scanner, activates those patches selected to receive, say, G. After exposure to light, the array is then flooded with G. Then the array is exposed to a different pattern and again flooded (with a different base), and so on. This technology is practicable up to about 20 cycles and is highly appropriate wherever linear heterooligomers sharing a common chemistry are required.
2. For all other cases, minute amounts of the receptor substances are directly deposited on the array (e.g., using laboratory microrobotics combined with inkjet technology for applying solutions of the different substances). This is suitable for

---

[2] If one is trying to determine whether certain genes are present in a bacterial culture (for example), the array would be coated with patches of complementary nucleic acid sequences. The DNA is extracted from the bacteria, subjected to some rudimentary purification, separated into single strands, and usually cut into fragments with restriction enzymes before pouring over the microarray.

[3] See Chumakov et al. (2005) for a discussion of design principles.

[4] Fodor et al. (1991).

large macromolecules, such as proteins, or sets of molecules of substances not sharing a common chemistry, or longer oligopeptides.

In both cases, each patch can be uniquely identified by its Cartesian array coordinate.

Sample preparation. The raw material is processed to make available the analyte(s) of interest and possibly partially purified. The mRNA is typically used to generate a set of complementary DNA molecules (cDNA), which may be tagged (labelled) with a fluorescent or other kind of label.

Array exposure. The array is flooded with the sample and allowed to reach equilibrium. Then all unbound sample is washed away. If the analyte was not tagged, tagging can be carried out now on the chip (e.g., by flooding with a hybridization-specific dye[5]) after removing the unbound molecules, which has the advantage of eliminating the possibility of the tag interfering with the binding.

Array reading. The array is scanned to determine which patches have captured molecules from the sample. If the sample molecules have been tagged with a fluorophore, then fluorescent patches indicate binding, with the intensity of fluorescence giving some indication of the amount of material bound, which, in turn, should be proportional to the amount of mRNA present in the original sample.

Image processing. The main task is to normalize the fluorescent (or other) intensities. Normalization is important when comparing the transcriptomes from two samples (e.g., taken from the same tissue subject to two different growth conditions). A straightforward way of achieving this is to assume that the total amount of expressed mRNA is the same in both cases (which may not be warranted, of course) and to divide the intensity of each individual spot by the sum of all intensities. If the transcriptomes have been labelled with different fluorophores and exposed simultaneously to the same chip, then normalization corrects for differences in fluorescence quantum yields and the like.

Analysis. The procedures followed for supervised hypothesis testing will depend on the details of the hypothesis (Sect. 8.2). Very commonly, unsupervised exploratory analysis of the results is carried out which, in effect, uses no prior knowledge but explores the data on the basis of correlations and similarities. One goal is to find groups of genes that have correlated expression profiles,[6] from which it might be inferred that they participate in the same biological process. Another goal is to group tissues according to their gene expression profiles; it might be inferred that tissues with the same or similar expression profile belong to the same clinical state.

If a set of experiments comprising samples prepared from cells grown under m different conditions has been carried out, then the set of normalized intensities (i.e.,

---

[5]E.g. ethidium bromide, the fluorescence of which becomes about 20-fold stronger after it is intercalated into double-stranded DNA.

[6]An expression profile is defined as a two-column table, with conditions in the left-hand column and the corresponding (relative) amounts of expressed proteins (possibly as RNA) in the right-hand column.

transcript abundances) for each experiment defines a point in m-dimensional expression space, whose coordinates give the (normalized) degrees of expression. Distances between the points can be calculated by, for example, the Euclidean distance metric, that is,

$$d = \left[\sum_{i=1}^{m} (a_i - b_i)^2 \right]^{1/2} , \tag{14.1}$$

for two samples a and b subjected to m different conditions. Clustering algorithms (Sect. 8.3.1) can then be used to group transcripts on the basis of their similarities. The hierarchical clustering procedure is the same as that used to construct phylogenies (Sect. 13.8); that is, the closest pair of transcripts forms the first cluster, the transcript with the closest mean distance to the first cluster forms the second cluster, and so on. This is the unweighted pair-group method average (UPGMA); variants include single-linkage clustering, in which the distance between two clusters is calculated as the minimum distance between any members of the two clusters, and so on.

Fuzzy clustering algorithms may be more successful than the above "hard" schemes for large and complex datasets. Fuzzy schemes allow points to belong to more than one cluster. Degree of membership is defined by

$$u_{r,s} = 1 / \sum_{j=1}^{m} \left(\frac{d(x_r, \theta_s)}{d(x_r, \theta_j)} \right)^{1/(q-1)} , r = 1, \ldots, N; \ s = 1, \ldots, m , \tag{14.2}$$

for N points and m clusters (m is given at the start of the algorithm), where $d(x_i, \theta_j)$ is the distance between the point x_i and the cluster represented by θ_j, and $q > 1$ is the fuzzifying parameter. The cost function

$$\sum_{i=1}^{N} \sum_{j=1}^{m} u_{r,s}^{j} d(x_i, \theta_j) \tag{14.3}$$

is minimized (subject to the condition that the $u_{i,j}$ sum to unity) and clustering converges to cluster centres corresponding to local minima or saddle points of the cost function. The procedure is typically repeated for increasing numbers of clusters until some criterion for clustering quality becomes stable; for example, the partition coefficient

$$(1/N) \sum_{i=1}^{N} \sum_{j=1}^{m} u_{i,j}^{2} . \tag{14.4}$$

The closer the partition coefficient is to unity, the "harder" (i.e., the better separated) the clustering.

Instead of using a clustering approach, the dimensionality of expression space can be reduced by principal component analysis (PCA), in which the original dataset is projected onto a small number of orthogonal axes. The original axes are rotated until there is maximum variation of the points along one direction. This becomes the first principal component. The second is the axis along which there is maximal residual variation, and so on (see also Sect. 8.3.2).

Limitations and alternatives

Microarrays have some limitations, and one should note the following potential sources of problems: manufacturing reproducibility; variation in how the experiments are carried out [exposure duration (is equilibrium reached?), temperature gradients, flow conditions, and so on, all of which may severely affect the actual amounts hybridized]; ambiguity between preprocessed and postprocessed (spliced) mRNA; mRNA fragment size distribution not matching that of the probes; quantitative interpretation of the data; expense. Attempts are being made to introduce globally uniform standards—minimum information about a microarray experiment (MIAME)—in order to make comparison between different experiments possible. Other techniques have been developed, such as serial analysis of gene expression (SAGE). In this technique, a short but unique sequence tag is generated from the mRNA of each gene using PCR (Sect. 13.1.2) and joined together ("concatemerized"). The concatemer is then sequenced. The degree of representation of each tag in the sequence will be proportional to the degree of gene expression.

The transcription products of many closely related genes such as those originating from alternative mRNA splicing (Sect. 10.7.5) may be difficult to distinguish using standard microarray techniques; efforts to overcome that problem include the use of bundles of tens of thousands of of optical fibres, to the ends of which thousands of glass beads, each loaded with a particular DNA sequence, are fixed.[7] Since the beads are comparable in size (a few micrometres in diameter) with the optical fibre cores each fibre will carry at most one bead. Each fibre is individually addressable and the DNA sequence associated with it is first identified using fluorescent complementary DNA fragments. The attraction of the technique is the enhanced sensitivity.

Problem. How many n-mers are needed to unambiguously identify g genes?

14.2   Proteomics

The proteome can be accessed directly by measuring the expression levels, not of the mRNA transcripts but of the proteins into which they are translated. Not surprisingly, in the relatively few cases for which comparative data for both the transcriptome and proteome have been obtained, the amounts of the RNAs and corresponding proteins may be very different, even if all the different proteins derived from the same RNA are grouped together—translation is an important arena for regulating protein

---

[7] Yeakley et al. (2002) These researchers combined their fibre optic array with the technique of RNA-mediated annealing, selection and ligation (RASL), in which the mRNAs produced in a particular cell type are extracted and mixed with DNA oligomers whose sequences are complementary to those at which two RNA sections could be joined by splicing ("splice junctions"); the presence of a particular splice junction leads to binding of the DNA oligomers, which can then be multiplied, fluorescently labelled and exposed to the optical fibre array with which the sequences can be identified.

synthesis. Before this became apparent, transcriptomics acquired importance because technically it is much easier to obtain the transcriptome using a microarray than it is to obtain the proteome using laborious two-dimensional gel electrophoresis (Sect. 14.2.1), for example. It was hoped that the transcriptome would be a reasonably faithful mirror of the proteome. This is, however, definitely not the case in general; there is no presently discernible unique relationship between the abundance of mRNA and the abundance of the corresponding protein. Hence, the transcriptome has lost some of its importance; it is "merely" an intermediate stage and does not contribute directly to phenotype in the way that the proteome does. Furthermore, the transcriptome contains no information about the very numerous post-translational modifications of proteins. On the other hand, to understand the relation between transcriptome and proteome would be a considerable advance in understanding the overall mechanism of the living cell. At present, given that both transcriptome and proteome spaces each have such a high dimensionality, deducing a relation between trajectories in each is a rather forlorn hope.

The first step in proteomics proper is to separate all of the expressed proteins from each other such that they can be individually quantified (i.e., characterized by type and number). Prior to that, however, the ensemble of proteins have to be separated from the rest of the cellular components. Cells are lysed, proteins are solubilized, and cellular debris is centrifuged down. Nucleic acids and lipids are removed and sometimes very abundant proteins (such as albumin from serum). A subset of proteins may be labelled at this stage, to assist later identification.

A particularly useful form of labelling is to briefly (for 30–40 min) feed the living cells with radioactive amino acids (^{35}S-cysteine and methionine are suitable), followed by an abundance of nonradioactive amino acids. The degree of incorporation of radioactivity into the proteins is then proportional to the net rate of synthesis (i.e., biosynthesis rate minus degradation rate).

The two main techniques for separating the proteins in this complex mixture (which is likely to contain several hundred to several thousand different proteins) are the following:

1. Two-dimensional gel electrophoresis (2DGE);
2. Enzymatic proteolysis into shorter peptides followed by column chromatography. Trypsin is usually used as the proteolytic enzyme (protease) since it cuts at well-defined positions (lysines).

The protein mixture may be pretreated (prefractionated), using chromatography or electrophoresis, before proceeding to the separation step, in order to selectively enrich it with certain types of proteins.

Problem. List and discuss the differences between mRNA and protein abundances.

14.2.1 Two-Dimensional Gel Electrophoresis

In order to understand the principles of protein separation by 2DGE, let us first recall some of the physicochemical attributes of proteins. Two important ones are:

1. Molecular weight M_r (relative molecular mass);
2. Net electrostatic charge Z [as a function of pH—the pH at which $Z = 0$ is important as a characteristic parameter, known as the isoelectric point (i.e.p.), or pI, or point of zero charge (p.z.c.)].

Both can be calculated from the amino acid sequence (assuming no post-translational modifications), provided M_r and Z of the individual amino acids are known. M_r is easy; to calculate Z, one has to make the quite reliable assumption that all of the ionizable residues are on the protein surface. The calculation is not quite as simple as adding up all the surface charges, since they mutually affect each other (cf. the surface of a silicate mineral: not every hydroxyl group is ionized, even at extremely low pH).[8]

The technique itself was developed by Klose and, independently, by O'Farrell in 1975. The concept depends on the fact that separation by isoelectric point (i.e.p.) is insufficient to separate such a large number of proteins, many of whose i.e.p.s are clustered together. Equally, there are many proteins with similar molecular masses. By applying the two techniques sequentially, however, they can be separated, especially if large (30×40 cm) gels are used.

Proteins in the crude cell extract are dispersed in an aqueous medium containing the anionic detergent sodium dodecyl sulphate (SDS), which breaks all noncovalent bonds (i.e., subunits are dissociated, and probably denatured too); the first separation takes place according to the i.e.p. by electrophoresis on a gel along which a pH gradient has been established; the partly separated proteins are then transferred to a second, polyacrylamide, gel within which separation is effected according to size (i.e., molecular weight if all proteins are assumed to have the same density).

If the cells have been pulse radiolabelled prior to making the extract, then the final gel can be scanned autoradiographically and the density of each spot is proportional to the net rate of protein synthesis. Alternatively (or in parallel) the proteins can be stained and the gel scanned with a densitometer; the spot density is then proportional to protein abundance. There are some caveats: Membrane proteins with more than two transmembrane sequences are poorly recovered by the technique; if ^{35}S met/cys is used, one should note that not all proteins contain the same number of met and cys (but this number is only very weakly correlated with molecular weight); autoradiography may underestimate the density of weak spots, due to low-intensity

---

[8]Linderstrøm-Lang worked out a method of taking these correlations into account; his formula works practically as well as more sophisticated approaches (including explicit numerical simulation by Brownian dynamics; cf. Madura et al. 1994) and is much simpler and more convenient to calculate (see Ramsden et al. 1995 for an application example).

Fig. 14.2 A two-dimensional gel after staining

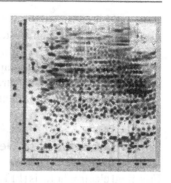

reciprocity failure of the photographic (silver halide) film used to record the presence of the radionucleides; the commonly used Coomassie blue does not stain all proteins evenly, although the unevenness appears to be random and hence should not impose any systematic distortion on the data; rare proteins may not be detected at all; several abundant proteins clustered close together may not be distinguishable from each other; and very small and very large proteins, and those with isoelectric points (pI) at the extremes of the pH range, will not be properly separated. The molecular weight and isoelectric point ranges are limited by practical considerations. Typical ranges are $15\,000 < M_r < 90\,000$ and $3 < pI < 8$. Hence, the mostly basic (pI typically in the range 10–14) 50–70 ribosomal proteins will not be captured, as a notable example (on the other hand, these proteins are not supposed to vary much from cell to cell, regardless of conditions, since they are essential proteins for all cells; hence, they are not considered to be especially characteristic of a particular cell or metabolic state). Figure 14.2 shows a typical result.

14.2.2 Column Chromatography

The principle of this method is to functionalize a stationary solid phase (granules of silica, for example) packed in a column and pass the sample (suspended or dissolved in the liquid mobile phase) through it (cf. Sect. 16.5.1). The functionalization is such that the proteins of interest are bound to the granules, and everything else passes through. A change in the liquid phase composition then releases the bound proteins. Better separations can be achieved by "multidimensional" liquid chromatography (MDLC), in which a cation exchange column (for example) is followed by a reverse phase column. The number of "dimensions" can obviously be further increased. Usually, the technique is used to prepurify a sample, but, in principle, using differential elution (i.e., many proteins of interest are bound and then released sequentially by slowly increasing pH or polarity of the liquid), high-resolution analytical separations may also be accomplished. Miniaturization (nano-liquid chromatography) offers promise in this regard. The output from the chromatography may be fed directly into a mass spectrometer (MS).

In multidimensional protein identification technology (MudPIT), the proteins are first denatured and their cysteines reduced and alkylated, and then digested with a protease. Following acidification, the sample is then passed through a strong cation exchange chromatographic column, followed by reverse phase chromatography. Eluted peptides are introduced into a mass spectrometer (typically a tandem (MS/MS) instrument) for identification (see Sect. 14.3).

14.2.3 Other Kinds of Electrophoresis

Free fluid electrophoresis (FFE) is distinguished from chromatography in that there is no stationary phase (i.e., no transport of analytes through a solid matrix such as a gel). The separation medium and the analytes are carried between electrodes, arranged such that the electric field is orthogonal to the flow of the separation medium.[9]

14.3 Protein Identification

Two-dimensional gel electrophoresis is very convenient since it creates a physical map of the cell's proteins in M_r–i.e.p. space, from which the proteins at given coordinates can actually be cut out and analysed. Hence, it is possible to apply Edman sequencing,[10] at least to the more abundant proteins, or Stark C-terminal degradation. The most widely applied technique is based on MS, however.[11] It is capable of much higher throughput, and post-translational modifications can be readily detected. Mass spectrometers consist of an ion source, a mass analyser (ion trap, quadrupole, time of flight (ToF), or ion cyclotron) and a detector

The objects to be analysed have to be introduced into the MS in the gas phase. This can be achieved by electrospraying or laser desorption ionization. In electrospraying, the proteins are dissolved in salt-free water, typically containing some organic solvent, and forced to emerge as droplets from the end of an electrostatically charged silica capillary. As the solvent evaporates, the electrostatic charge density increases until the droplets explode. The solution dilution should be such that each protein is then isolated from its congeners. The remaining solvent evaporates and the protein molecules pass into the MS. At this stage, each protein molecule is typically multiply charged. Sequential quadrupole filters, typically three, are used to achieve adequate discrimination. The mass spectrum for an individual protein consists of a series of

---

[9]See Patel and Weber (2003) for a review.

[10]The N-terminal of the protein is derivatized with phenylisothiocyanate to form a phenylthiocarbamate peptide, and the first amino acid is cleaved by strong acid resulting in its anilothiazolinone derivative plus the protein minus its first N-terminal amino acid. The anilothiazolinone derivative is converted to the stabler phenylthiohydantoin for subsequent high-performance liquid chromatography (HPLC) identification.

[11]See Bell et al. (2009) for efforts to overcome errors in mass spectrometry-based proteomics.

peaks corresponding to m/z ratios whose charge z differs by one electron. The middle quadrupole may contain a collision gas (e.g., argon) to fragment the protein into smaller peptides.

In laser desorption ionization, usually called matrix-assisted laser desorption ionization (MALDI) or surface-enhanced laser desorption/ionization (SELDI), the protein is mixed with an aromatic organic molecule [e.g., sinapinic acid $((CH_3O)_2 OHC_6H_2(CH_2)_2COOH)$] spread out as a thin film, and irradiated by a pulsed ultraviolet laser. The sinapinic acid absorbs the light and evaporates, taking the proteins with it. Other matrices can be used with infrared lasers.[12] The proteins are typically singly charged, and a ToF MS detects all the ions according to their mass. MALDI-ToF MS cannot detect as wide a range of proteins as quadrupole MS, and the matrix can exert unpredictable effects on the results. Nevertheless, the vision of spots on a two-dimensional gel being rapidly and sequentially vaporized by a scanning laser and immediately analysed in the MS offers hope for the development of high-throughput proteomics analysis tools.

Newer developments in the field include the application of sophisticated ion cyclotron resonance MSs, the use of Fourier transform techniques, and miniature instrumentation according to the lab-on-a-chip concept.

Mass spectrometry is also used to characterize the peptide fragments resulting from proteolysis followed by chromatography. Proteins separated by 2DGE can also be cleaved using trypsin or another protease to yield fragments, which are then mass-fingerprinted using MS. The proteolytic peptide fragments are encoded as a set of numbers corresponding to their masses, and these numbers are compared with a database assembled from the mass-fingerprints from known peptides.

14.4 Isotope-Coded Affinity Tags

Isotope-coded affinity tags (ICATs)[13] are particularly useful for comparing the expression levels of proteins in samples from two different sources (e.g., cells before and after treatment with a chemical). It is a way of reducing the variety (number of proteins that have to be separated) of a complex mixture. Proteins from the two sources are reacted with light and heavy ICAT reagents in the presence of a reducing agent. The reagents comprise a biotin moiety, a sulfhydryl-specific iodoacetate moiety, and a linker that carries eight ^{1}H (light) or ^{2}H (heavy) atoms. They specifically tag cysteinyl residues on the proteins. The two batches are then mixed and the proteins cleaved using trypsin. The fragments, only about a fifth of which contain cysteine, can be readily separated by chromatography on an avidin affinity column (which binds to the biotin), and finally analysed by MS. Singly charged peptides of identical sequences from the two sources are easily recognized as pairs differing

---

[12]See Chem. Rev. 103 (2003), issue no 2.
[13]Developed by Aebersold (see Gygi et al. 1999).

by eight atomic mass units. Differences in their expression levels can be sensitively compared and normalized to correct for differences in overall protein content.

Many other affinity enrichment techniques can be imagined, tailored according to the proteins of interest; for example, lectins can be used to make a column selectively capturing glycoproteins.

14.5 Protein Microarrays

Generic aspects of microarrays have already been covered in Sect. 14.1. Protein microarrays allow the simultaneous assessment of expression levels for thousands of genes across various treatment conditions and time. The main difference compared with nucleic acid arrays is the difficulty and expense of placing thousands of protein capture agents on the array. Since capture does not depend on simple hybridization, but on a certain arrangement of amino acids in three-dimensional space, complete receptor proteins such as antibodies have to be used, and then there is the danger that their conformation is altered by immobilization on the chip surface.[14] It may be possible to exploit the advantages of nucleic acid immobilization (especially the convenient photofabrication method) by using aptamers—oligonucleotides binding specifically to proteins—for protein capture (this might be especially useful for determining the expression levels of transcription factors).

An ingenious approach is to prepare an array of genes (which is much easier than preparing an array of proteins) and then expose the microarray to a suitable mixture of *in vitro* transcription and translation factors (e.g., from reticulocytes), such that the proteins are synthesized *in situ*.[15]

Polypeptide immobilization chemistries typically make use of covalently linking peptide side chain amines or carboxyl groups with appropriately modified chip surfaces. Quite a variety of possible reactions exist, but usually several different residues are able to react with the surface, making orientational specificity difficult to achieve. Proteins recombinantly expressed with a terminal oligohistidine chain can be bound to surface-immobilized nickel ions, but the binding is relatively unstable.

A significant problem with protein microarrays is the nonspecific adsorption of proteins. Unfavourably oriented bound proteins, and exposed substratum, offer

---

[14] As an alternative way to prepare oligopeptide receptors, the phage display technique invented by Dyax is very useful. The gene for the coat protein expressed abundantly on the surface of a bacteriophage virus is modified by adding a short sequence coding for an oligopeptide to one end. Typically, a large number ($\sim 10^9$) of random oligonucleotides are synthesized and incorporated (one per phage) into the virus gene. The phages are then allowed to multiply by infecting a host bacterium; the random peptide is expressed in abundance on the coat of the phage along with the regular coat protein. The phage population is then exposed to an immobilized target (e.g., a protein). Any phage (a single one suffices) whose peptide interacts with the target during this screening is retained and recovered, and then multiplied *ad libitum* in bacteria.

[15] See Oh et al. (2007) for an example of this kind of approach.

targets for nonspecific adsorption. Pretreatment with a so-called "blocking" protein (seralbumin is a popular choice) is supposed to eliminate the nonspecific adsorption sites, although some interference with specific binding may also result.

As with the transcriptome, statistical analyses of protein microarray data focus on either finding similarity of gene expression profiles (e.g., clustering) or calculating the changes (ratios) between control and treated samples (differential expression).

14.6   Protein Expression Patterns—Temporal and Spatial

Whether the transcriptome or the proteome is measured, the result from each experiment is a list of expressed objects (mRNAs or proteins) and their abundances or net rates of synthesis. These abundances are usually normalized so that their sum is unity. Each experiment is therefore represented by a point in protein (or mRNA) space (whose dimension is the number of proteins; the distance along each axis is proportional to abundance); each protein is represented by a point in expression space (whose dimension is the number of experiments). The difficulty in making sense of these data is their sheer extent: There are hundreds or thousands of proteins and there may be dozens of experiments (which could, for example, be successive epochs in a growth experiment, or a series of shocks). Hence, there is a great need for drastic data reduction.

One approach has already been mentioned [Sect. 14.1; viz. to group proteins into blocks whose expression tends to vary in the same way (increase, decrease, remain unchanged)]. This is the foundation for understanding how genes are linked together into networks, as will be discussed in the next chapter.

Another approach is to search for global parameters characterizing the proteome. Considering it as "vocabulary" transferring information from genotype to phenotype, it has been found that the distribution of protein abundances p_r follows the same canonical law as the frequency of words in literary texts:[16]

$$p_r = P(r + \rho)^{-1/\theta} , r = 1, 2, \ldots, R \tag{14.5}$$

where the two parameters are the informational temperature θ, which is low for limited expression of the potential gene repertoire and high for extensive expression, and the effective redundancy ρ, which is high when many alternative pathways are active, and low otherwise. R is the total number of proteins that can be synthesized in the cell and P is a normalizing coefficient chosen such that the p_r sum to unity; they are ordered according to decreasing magnitude and r is the rank in this list.

Equation (14.5) is well suited to following the evolution of, for example, a synchronized culture of cells: the parameters are determined for each epoch sampled. The spatially differentiated expression of proteins is much more difficult to follow experimentally. The most effective technique is probably time of flight secondary

---

[16]See Vohradský and Ramsden (2001) and Ramsden and Vohradský (1998).

ion mass spectrometry (Tof-SIMS): secondary ions are sputtered from the sample by a primary iron gun firing, typically, gallium or oxygen ions. The secondary ions are extracted into the flight tube of a mass spectrometer and quantified.[17] One useful way to exploit this technology is to feed the organism under investigation with isotopically labelled food (^{15}N is a popular choice since it is not radioactive and the slight difference in mass from the naturally far more abundant ^{14}N is likely to have a negligible physiological effect). Any molecules containing ^{15}N will give a clear signature in the mass spectrometer. The spatial resolution is determined, *inter alia* by the fineness of the collimated primary ion beam.

14.7   The Kinome

One of the most fundamental mechanisms for reversible enzyme activation is phosphorylation. This reaction is catalysed by enzymes generically called kinases. Several hundred human kinases are known; collectively they comprise the kinome. Most commonly, serine or threonine residues are phosphorylated, but also tyrosine, histidine and others are known. The so-called mitogen-activated protein kinases (MAPK), including MAPK kinases (MAPKK) and MAPKK kinases, comprise perhaps the best known family.[18] Phosphorylation introduces a bulky, negatively charged (at neutral pH) group into the amino acid. These changes in both the size and the charge of the residue typically induce significant conformational changes in the protein; it is easy to understand in these general terms how phosphorylation of an enzyme (which might itself be a kinase) can have a profound impact on its activity: typically, phosphorylation activates an enzyme that is otherwise catalytically inert. The reverse reaction, dephosphorylation, is carried out by enzymes called phosphatases.[19]

The propagation of the signal can be described by a hidden Markov model (Sect. 13.5.2). Let the substrate of a kinase (e.g., MAPK) be denoted by X and the phosphorylated substrate by XP. When a kinase is in its resting, inactive form, the following would be a reasonable guess at the transition probabilities:

$$\begin{array}{c|cc} \rightarrow & X & XP \\ \hline X & 1.0 & 0.0 \\ XP & 0.9 & 0.1 \end{array} \tag{14.6}$$

since the phosphatases are permanently active. However, if the MAPK is itself phosphorylated, the transition probabilities change:

$$\begin{array}{c|cc} \rightarrow & X & XP \\ \hline X & 0.0 & 1.0 \\ XP & 0.9 & 0.1 \end{array}. \tag{14.7}$$

---

[17]Fearn (2015).

[18]See, e.g., Kolch et al. (2005).

[19]See Johnson and Hunter (2005) for a review of experimental methods for determining phosphorylation.

The phosphorylation of the MAPK itself can be represented as a Markov chain and, if X is itself a kinase, the transition probabilities for the phosphorylation of *its* substrate will also be different for X and XP. The necessity of the phosphatases (whose effect is represented by the transition probability $p_{XP \to X}$) is clearly apparent from this scheme, for without them the supply of substrate would be quickly exhausted.

An insidious form of toxicity is engendered by molecules capable of more or less indiscriminate phosphorylation of proteins. The extreme importance of phosphorylation for, especially, intracellular communication suggests that if such molecules penetrate into the body they are likely to wreak havoc on intracellular signalling. As already mentioned, the dephosphorylating phosphatases are permanently active, but some of the anomalously phosphorylated molecules might not be substrates for them. Furthermore, the phosphorylation takes time and if the upset is great enough, the damage might become irreversible. Organophosphorus compounds causing organophosphate-induced delayed neuropathy (such as some tricresyl phosphate isomers, diisopropyl fluorophosphate, a nerve gas, and Mipafox, an insecticide) are candidates for the disruptive phosphorylation of kinases.[20]

14.8 Biochemical Signalling

The organization of kinases into signalling cascades, in which a phosphorylated, hence activated, enzyme itself phosphorylates and activates another kinase, is characteristic. One of the consequences of such cascades is great amplification of the initial signal (which might have been a single molecule). This is a robust method for overcoming noise (cf. Sect. 3.6). A cascade also achieves fanout, familiar to the designer of digital electronic circuits,[21] in which an output is made available to multiple devices. If the response to the external stimulus triggering the cascade requires the activation of multiple enzymes, for which the genes encoding them might be located on different chromosomes, the cascade is a way of achieving rapid diffusion of the signal in a relatively unstructured milieu. Furthermore, as a protein, each element of the cascade is only able to interact with a relatively small number of other molecules bearing information.[22] There may be more potentially blocking molecules than sites on a single member of the cascade. The blocking effect can be achieved by interacting with any cascade member, since the entire cascade essentially constitutes a single linear channel for information flow.

---

[20]Lapadula et al. (1992).

[21]And, indeed, to the neurologist—see Chap. 17.

[22]This limitation is imposed physicochemically; for example, there is only room for a small number of other proteins to cluster round and interact with a central one and, of course, the entire surface of the central protein is unlikely to be sensitive to the presence of other proteins; the possibilities for interaction are typically limited to a small number of specific binding sites.

The fact that information is conveyed by material objects, whose supply is variable and limited and which occupy an appreciable proportion of the volume of the cell, creates a situation that is significantly different from that of regulatory networks based on fixed (e.g., electrical or optical) connexions. As was already stressed in the discussion of the kinase-based signalling pathways (Sect. 14.7), the information-bearing "quanta" have to be regenerated by phosphatases. There is, moreover, an ultimate constraint in the form of the finiteness of the attributes of a cell; conceivably, it could happen that all of the kinases were converted to the active form and no resources were available for regenerating them, and hence no resources for communicating the need for regeneration.

Problem. Describe some examples of signalling cascades (e.g., blood clotting, glycogenolysis).

Problem. Using the formalism of Figs. 3.1 and 9.2 analyse one of the cascades described in the previous problem, or the MAPK signalling system, from an information-theoretic viewpoint.

References

Bell AW et al (2009) A HUPO test sample study reveals common problems in mass spectrometry-based proteomics. Nat Methods 6:423–430

Chumakov S, Belapurkar C, Putonti C et al (2005) The theoretical basis of universal identification systems for bacteria and viruses. J Biol Phys Chem 5:121–128

Fearn S (2015) Characterisation of biological material with ToF-SIMS: a review. Mater Sci Technol 31:148–161

Fodor SPA, Read JL, Pirrung MC et al (1991) Light-directed, spatially addressable parallel chemical synthesis. Science 251:767–773

Gygi SP, Rist B, Gerber SA et al (1999) Quantitative analysis of complex protein mixtures using isotope-coded affinity tags. Nat Biotechnol 17:994–999

Johnson SA, Hunter T (2005) Kinomics: methods for deciphering the kinome. Nat Methods 2:17–25

Kolch W, Calder M, Gilbert D (2005) When kinases meet mathematics. FEBS Lett 579:1891–1895

Lapadula ES, Lapadula DM, Abou-Donia MB (1992) Biochemical changes in sciatic nerve of hens treated with tri-o-cresyl phosphate: increased phosphorylation of cytoskeletal proteins. Neurochem Int 20:247–255

Madura JD, Davis ME, Gilson MK et al (1994) Biological applications of electrostatic calculations and Brownian dynamics simulations. Rev Comput Chem 5:229–267

Oh Y-H, Kim Y-P, Kim H-S (2007) SUMO chip for analysis of SUMO-conjugation to a target protein. Biochip J 1:28–34

Patel PD, Weber G (2003) Electrophoresis in free fluid: a review of technology and agrifood applications. J Biol Phys Chem 3:60–73

Ramsden JJ, Roush DJ, Gill DS et al (1995) Protein adsorption kinetics drastically altered by repositioning a single charge. J Am Chem Soc 117:8511–8516

Ramsden JJ, Vohradský J (1998) Zipf-like behavior in procaryotic protein expression. Phys Rev E 58:7777–7780

Vohradský J, Ramsden JJ (2001) Genome resource utilization during pro-caryotic development. FASEB J 15:2054–2056

Yeakley JM, Fan J-B, Doucet D, Luo L, Wickham E, Ye Z, Chee MS, Fu X-D (2002) Profiling alternative splicing on fiber-optic arrays. Nat Biotechnol 20:353–358

The Glycome, Lipidome and Microbiome

15

In comparison with the immense accumulations of knowledge in genomics and proteomics, data concerning these other -omes is still relatively sparse. Glycomics encompasses the polysaccharides of our body; lipidomics the repertoire of lipids; and microbiomics the resident bacterial community associated with our bodies.

15.1 Glycomics

Glycomics encompasses both the numerous glycans (polysaccharides) that exist independently and those that are conjugated—polysaccharides linked to proteins and lipids. By this means the variety of proteins and lipids can be enormously increased.

The study of the glycome is very much more complicated than the genome, proteome or lipidome, due to its enormous structural diversity, and global, high-throughput methods are still lacking although carbohydrate microarrays (cf. Sect. 14.5) are now being developed.[1] Some of the different classes requiring different experimental approaches include glycoproteins, glycolipids, N-glycans, O-glycans, neutral glycans, and sulfated (negatively charged) glycans. Mass spectrometry is the technique of choice for analysing glycan structure once they have been isolated.

15.2 Lipidomics

It is now known that the eukaryotic lipidome typically comprises many hundreds of different molecules, and their global analysis requires high-throughput techniques. An important development has been "shotgun" mass spectrometry of the lipids extracted by solvents,[2] which not only enables the different lipids to be identified, but

---

[1] Feizi et al. (2003).
[2] Ejsing et al. (2009).

© Springer-Verlag London 2015
J. Ramsden, *Bioinformatics*, Computational Biology 21,
DOI 10.1007/978-1-4471-6702-0_15

also quantifies their abundances. The high throughput is achieved by considerable automation of the process and the data handling is computationally heavy.[3]

15.3 Microbiomics

It is perhaps well known nowadays that the number of individual microbial cells in a typical human being exceeds the number of proper cells by about an order of magnitude. The importance of these guests in processing ingested nutrients can scarcely be overestimated, not only in our own species but also in, for example, ruminants, which could not otherwise digest cellulose. Until now, however, efforts to investigate this microbial richness seem to have been disproportionately small with respect to its evident importance for the maintenance of a healthy organism. A corollary of this relative neglect has been the rather cavalier attitude of present-day medicine towards this indispensable miniature ecosystem; in effect it is simply neglected. For example, antibiotics are frequently prescribed to eliminate a bacterial infection of the "main" organism, heedless of the devastation that the antibiotics are likely to inflict on gastrointestinal microbial diversity, with the likelihood of consequential ill health. Of similar concern is the well-nigh endemic use of certain herbicides, the residues of which remaining in food for human consumption may have deleterious effects on gut microbes.[4]

The vast increase in DNA sequencing capability has, of course, given a tremendous impetus to the study of our bacterial ecosystem, termed the microbiome (cf. Sect. 13.9), and it has been possible to initiate the study of how it varies temporally and spatially.[5] In future, such studies will doubtless be extended to fungi and viruses.

References

Costello EK, Lauber CL, Hamady M, Fierer N, Gordon JI, Knight R (2009) Bacterial community variation in human body habitats across space and time. Science 326:1694–1697

Ejsing CS, Sampaio JL, Surendranath V, Duchoslav E, Ekroos K, Klemm RW, Simons K, Shevchenko A (2009) Global analysis of the yeast lipidome by quantitative shotgun mass spectrometry. Proc Natl Acad Sci USA 106:2136–2141

Feizi T, Fazio F, Chai W, Wong C-H (2003) Carbohydrate microarrays—a new set of technologies at the frontiers of glycomics. Curr Opin Struct Biol 13:637–645

Samsel A, Seneff S (2013) Glyphosate's suppression of cytochrome P450 enzymes and amino acid biosynthesis by the gut microbiome: pathways to modern diseases. Entropy 15:1416–1463

Yetukuri L, Ekroos K, Vidal-Puig A, Orešič M (2008) Informatics and computational strategies for the study of lipids. Mol Biosyst 4:121–127

---

[3] Yetukuri et al. (2008).
[4] Samsel and Seneff (2013).
[5] Costello et al. (2009).

Interactomics: Interactions and Regulatory Networks

16

It is clear that the living cell (and *a fortiori* the multicellular organism) comprises a great variety of different components that must somehow be integrated into a functional whole. The framework of this integration is directive correlation (Fig. 9.1) and it may be considered as a problem of regulation.

Regulation was introduced in Chap. 9 (Sect. 9.4) as a means of ensuring that the system's output remained within its essential variables while its environment was undergoing change—in other words, as one of the mechanisms of adaptation (which is itself a special case of directive correlation). We are perhaps most familiar with regulation whereby the volition of the regulator is transformed into direct action—such as pressing the accelerator pedal of a motor car. In a steam locomotive, the lever with equivalent function is actually called the regulator. Stationary steam engines providing mechanical power to a factory or mine are typically required to run at a constant speed and are equipped with a "governor" (a device mounted on the spindle turned by the engine that increases its radius with increasing angular velocity of the spindle, due to centrifugal force and, via a system of cranks and levers, directly closes a valve shutting off steam to the driving cylinders) that automatically regulates the speed (this is another example of the "regulation by error" described in Sect. 9.4).

In these examples—and in numerous others in which the communication channels along which information flows are conducting wires carrying electrons—the elements constituting the regulated system are physically connected by levers, wires, or pipes. In the living cell, a *signal* (Sect. 14.8) is typically a transformed molecule, such as an activated enzyme (cf. Sect. 14.7), that simply diffuses away from where it is generated (cf. Sect. 6.3). Rather like certain male fish mating by merely dispersing their sperm in the water around them, to be picked up by any females of that species that happen to be in the vicinity, the transformed, information-bearing molecules will only catalyse the reaction for which they are activated if they encounter their specific substrate, to which they must first bind.[1] Hence, physicochemical affinities

---

[1] Eukaryotic cells in particular are in a great deal more structured than the simple picture suggests: Filaments of various kinds (e.g., microtubules) appear to function *inter alia* as tracks along which certain molecules are transported to specific destinations. However, even in this case, the

(interactions) between molecules play an essential rôle in regulation. From the base-pairing of nucleic acids, to the formation of the bilayer lipid membranes enclosing organelles and cells, through to the protein–protein interactions building up supramolecular complexes serving structural ends, or for carrying out reactions, the regulation of gene expression by transcription factors binding to promotors, the operation of the immune system—the list seems to be almost endless—one observes the molecules of life linked together in a web of interactions. The set of all these interactions (i.e., a list of all the molecules, associated with all the other molecules with which some kind of specific association is found) constitutes the interactome (the repertoire of interactions).[2]

If the proteins are considered as the nodes of a graph (cf. Sect. 7.2), a pair of proteins will be joined by a vertex if the proteins associate with each other. On this basis, the "interactome"—the set of interactions in which a protein could participate—would be characterized by such a graph, or an equivalent list of all the proteins in a cell, each associated with a sublist of the proteins with which they interact. This is in contrast to metabolic networks, in which two metabolites are joined if there is a chemical reaction (catalysed by an enzyme) leading from one to another (Sect. 18.4). Attention is often focused on small portions of these networks, which are then called pathways. The so-called signalling networks (Sect. 14.8) are of a similar nature, focusing on reactions such as protein phosphorylation to activate an enzyme (cf. Sect. 14.7), and they differ from metabolic networks only inasmuch as the enzyme substrates and reaction products are not metabolites, and the destination of many of the signalling pathways is typically a gene promoter site.

All proteins are, of course, gene products.[3] Hence, the fundamental regulatory network is that of the genes, which constitute the nodes, the edges signifying the activation or inhibition of other genes, and the central problem is to infer ("reverse engineer") both the state structure of the network (cf. Fig. 7.1) and the physical network of interactions. For the former, the input data are now typically the temporal evolution of gene expression profiles, obtained by a succession of microarray experiments (cf. Sect. 14.6). For the latter, association is measured more or less directly using a variety of physicochemical techniques. In this chapter, we will first deal with the problem of deducing the state structure from gene expression data and, in the second half, we will examine the physical interactions between molecules.

The graph of interactions is potentially extraordinarily large and complex. Even if one confines oneself to the N expressed proteins in a cell, there are $\sim N^2$ potential

---

(Footnote 1 continued)

information-bearing ("signalling") molecule has first to encounter, and bind to, the carrier molecule that will convey it along the track.

[2]McConkey has coined the term "quinary structure" (of proteins) for this web of interactions.

[3]This statement, the obvious corollary of the central dogma, is actually quite problematical—in the sense of having a rather ambiguous meaning—when scrutinized in detail. Many functionally relevant proteins are significantly modified (e.g., glycosylated) by enzymes after translation. Of course, these enzymes themselves are gene products.

binary interactions and vastly more higher-order ones.[4] Even if only a small fraction of these interactions actually occur (and some general results for the stability of systems (Sect. 7.1) suggest that only about 10 % will occur), we are still talking about $\sim 10^7$ interactions, assuming about 10^4 expressed proteins (in a eukaryotic cell), and 10^8 pairs would have to be screened in order to find the 10 %. In a prokaryote, with possibly only 1000 expressed proteins, the situation is more tractable but still poses a daunting experimental challenge, even without considering that many of those proteins are present in extremely low concentrations.

When one or more stimuli arrive at a cell, the affinities of certain proteins for a transcription factor binding site (TFBS) are altered, and mRNA transcription is activated or inhibited, resulting in altered abundance of the mRNA and the translated protein, measured using microarrays (Sect. 14.1). To a first approximation, it is useful to represent expression as "1" and the absence of expression as "0". Alternatively, since many proteins are nearly always expressed to some extent, increased transcription–translation ("upregulation") can be represented as "1", and decreased transcription–translation ("downregulation") as "0". The system can then be analysed as a Boolean network.

In prokaryotes, and possibly some eukaryotes, genes are organized in operons. As already discussed in Chap. 10, an operon comprises a promoter sequence controlling the expression of several genes (positioned successively dowstream from the promoter), whose products may be successive enzymes in a metabolic pathway.[5] In most of the eukaryotes investigated hitherto, a similar but less clearly delineated arrangement also exists: The same transcription factor may control the expression of several genes, which may, however, be quite distant from each other along the DNA, indeed even on different chromosomes.

Genes observed to be close to each other in expression space are likely to be controlled by the same activator. Each gene can have its own promoter sequence; coexpression is then achieved by the transcription factor binding to a multiplicity of sites. Indeed, given that several factors may have to bind simultaneously to the TFBS region in order to modulate expression, control appears to be most commonly of the "many to many" variety, as anticipated many years ago by Wright. Since genes code for proteins, which, in turn, control the expression of other genes, the network is potentially extremely interconnected and heterarchical.

Example. The *lac* operon (part of the DNA of *E. coli*) consists of consecutive repressor gene, promoter, operator, and lactose-metabolizing gene sequences. In the absence of lactose, the repressor protein binds to the operator sequence and prevents the RNA polymerase from transcribing the genes (of which there are three, translated into permease, a protein that helps to transports lactose into the cell, and β-galactosidase, and galactoside transacetylase). Allolactose, a by-product of lactose metabolism, is able to bind to the repressor, changing its conformation and

---

[4]Many transcription factors, for example, are multiprotein complexes.
[5]Groups of operons controlled by a single transcription factor are called regulons; groups of regulons are called modulons.

preventing it from binding to the operator sequence, whereupon the RNA polymerase is no longer prevented from binding to the promoter sequence and hence initiates transcription of the lactose-metabolizing genes. Note that a certain basal level of production of the lactose-metabolizing proteins is necessary.

Problem. Construct a Boolean model of the *lac* operon. *Hint*: Start with a very simple model and progressively add features. Can the effects of noise and delays in signal transmission be incorporated?

Each gene will have its experimentally determined expression profile, and once these data are available, the genes can be clustered (Sect. 8.3.1) or arranged into a hierarchy (Sect. 13.8). The principal task, however, is to deduce the state structure from such data.

It is a very useful simplification to consider the model networks to be Boolean (i.e., genes are switched either on or off). To give a flavour of the approach, consider an imaginary mini-network in which gene A activates the expression of B, B activates A and C, and C inhibits A.[6] This is just an abbreviated way of saying that the translated transcript of A binds to the promoter sequence of B and activates transcription of B, and so on. Hence, A, B, and C form a network, which can be represented by a diagram of immediate effects (cf. Fig. 9.2) or as a Boolean weight matrix:

$$
\begin{array}{c|ccc}
 & A & B & C \\
\hline
A & 0 & 1 & -1 \\
B & 1 & 0 & 0 \\
C & 0 & 1 & 0
\end{array}
\tag{16.1}
$$

Reading from top to bottom gives the cybernetic formalization; reading horizontally gives the Boolean rules: A = B NOT C, B = A, C = B. Matrix (16.1) can be transformed to produce a stochastic matrix (a probabilistic Boolean network) and the evolution of transcription given by a Markov chain. Different external circumstances engendering different metabolic pathways can be represented by hidden Markov models (Sect. 13.5.2). Noise can be added in the form of a random fluctuation term. Alternatively, the system can be modelled as a neural net in which the evolution of the expression level a_i (i.e., the number of copies produced) of the ith protein in time τ is

$$
\tau \frac{\mathrm{d}a_i}{\mathrm{d}t} = \mathcal{F}_i \left(\sum_j w_{ij} a_j - x_i \right) - a_i ,
\tag{16.2}
$$

where w is an element of the weight matrix (16.1), $\mathcal{F}$ is a nonlinear transfer function (e.g., an exponential function), x is an external input (e.g., a delay), and the negative term at the extreme right represents degradation. The Boolean network approach lends itself to elegant, compact descriptions that can easily be extended to hundreds of genes.

---

[6]After Vohradský (2001).

16.1 Inference of Regulatory Networks

Given the experimental microarray data consisting of g gene transcripts measured at t successive epochs, one seeks to find how expression is controlled by a relatively small number $c \ll g$ of control nodes, represented as an $g \times c$ matrix R. This implies decomposition of the experimental $g \times t$ matrix E:

$$E = RF \tag{16.3}$$

where F is a $c \times t$ matrix giving the temporal evolution of the control nodes. However, this decomposition is not, in general, unique. Inference of the network is still largely a heuristic procedure, in which alternative topologies fitting the data equally well are considered, and, finally, a selection is made on the basis of additional, *ad hoc*, information. The field of systems biology[7] is largely devoted to this problem.

Many new developments are under way. Petri nets may be able to incorporate more biological features while still retaining a compact description. Representing network components as tensors allows many standard manipulations to be carried out, some of which may turn out to be useful in revealing useful features of the data. For more complete quantification, explicit differential equations for regulation are more successful,[8] in which the temporal variation of expression of a gene product z under the effect of m regulators is written as

$$\frac{dz}{dt} = \frac{k_1}{1 + \exp\left(-\sum_{j=1}^{m} w_j y_j(t) + b\right)} - k_2 z , \tag{16.4}$$

where k_1 is the maximum rate of expression, the y represent the expression levels of the regulators (usefully approximated as polynomials) and w are their rates, b represents delay, and k_2 is the rate coefficient for degradation of z. This system of equations can be fitted to experimental microarray data.

16.2 The Physical Chemistry of Interactions

Although knowledge of the state structure of a network (system) does not require knowledge of the physical structure, there can be no information transfer, and hence no regulatory control, in the absence of physical interaction. "Interaction", as implied by elementary chemical reactions of the type

$$A + B \underset{k_d}{\overset{k_a}{\rightleftharpoons}} C , \tag{16.5}$$

where C is a complex of A and B and for which an affinity (or equilibrium) constant K is defined according to the mass action law (MAL) by

$$K = \frac{ab}{c} , \tag{16.6}$$

---

[7] Kitano (2002).
[8] Vu and Vohradský (2007).

where the lowercase letters denote mole fractions,[9] is nearly always quite inadequate to characterize the association between two proteins. In practical terms, if an experiment is carried out with scant regard to the underlying physical chemistry, even slight differences in the way of carrying out the reaction or in the way the data are interpreted could result in considerable differences in the corresponding numerical values attributed to the interaction. At present, the interactome has mostly been assembled on the basis of dichotomous inquiry (i.e., does the protein interact or does it not?), but as technical capabilities improve, this is obviously going to change, and it will become important to assign gradations of affinity to the interactions.

The cytoplasm is crowded and compartmentalized. Hence, many pairs of proteins potentially able to interact have a negligible chance of encountering each other in practice. Moreover, local concentrations of inorganic ions and small molecules, which may greatly influence the strength of an interaction, often differ greatly from place to place within the cell. This gives an advantage to methods probing interactions *in vivo* over those requiring the proteins to be extracted. On the other hand, *in vivo* measurements cannot usually yield data sophisticated enough to go beyond the elementary model of interaction encapsulated by Eq. (16.5) and mostly cannot go beyond a simple yes/no appraisal of interaction. Additionally, unless the *in vivo* technique involves some three-dimensional spatial resolution, the result will be an average over different local microenvironments, physiological states, and so forth. On the other hand, properly designed *in vitro* experiments can reconstitute conditions of a tightly defined, spatially restricted physiological state of a living cell.

It should be emphasized that many protein interactions take place at the internal surfaces of cells, such as the various lipid bilayer membranes. The physical chemistry of the interactome is thus largely the physical chemistry of heterogeneous reactions, not homogeneous ones. It also follows that the interactions of the proteins with these internal surfaces must also be investigated: Clearly, a situation in which two potentially interacting partners become associated with a membrane, and then diffuse laterally until they encounter each other, is different from one in which only one protein is associated with the membrane, and the interacting partner remains in the bulk.

The field can naturally be extended to include the interactions of proteins with other nonprotein objects, such as DNA, RNA, oligosaccharides and polysaccharides, as well as lipid membranes. Indeed, it is essential to do so in order to obtain a proper representation of the working of a cell. Although the interactome emerged from a consideration of proteins, protein–DNA and protein–saccharide interactions are exceedingly important in the cell (the latter have been given comparatively less attention).[10]

One proposed simplification has been to consider that protein–protein binding takes place via a relatively small number of characteristic polypeptide domains (i.e.,

---

[9]In the literature, K is often loosely defined using Eq. (16.6) with concentrations rather than mole fractions, whereupon it loses its dimensionless quality.

[10]Remarkable specificity is achievable (see, e.g., Popescu and Misevic 1997).

a sequence of contiguous amino acids, sometimes referred to as a "module"). In the language of immunology, a binding module is an epitope (cf. Sect. 10.5). The module concept implies that the interactome could effectively be considerably reduced in size. There is, however, no consistent way of defining the modules. It seems clear that a sequence of contiguous amino acids is inadequate to do so; an approach built upon the dehydron concept[11] would appear to be required.

It is useful to consider two types of protein complexes: "permanent" and "transient". By permanent, large multiprotein complexes such as the spliceosome (and, in principle, any multisubunit protein) that remain intact during the lifetime of their constituents are meant. On the other hand, transient complexes form and disintegrate constantly as and when required. The interactome is thus a highly dynamic structure and this kinetic aspect needs to be included in any complete characterization.

The kinetic mass action law (KMAL) defines the same K as given in Eq. (16.6) according to

$$K = \frac{k_a}{k_d}, \qquad (16.7)$$

where the ks are the rate coefficients for association (a) and dissociation (d), but as it is a ratio, the same value of K results from association reactions that take either milliseconds or years to reach equilibrium. This temporal aspect can have profound influences on the outcome of a complex interaction. Many biological transformations (of the type often referred to as signal transduction) require the sustained presence of A in the vicinity of B in order to effect a change (e.g., of conformation) in B that will then trigger some further event (e.g., in C, also bound to B). A very well-characterized example of this kind of effect is the photolysis of silver halides.[12] Freshly reduced Ag will relax back to Ag^+ if it fails to capture another electron within a characteristic time (this is the origin of the low-intensity reciprocity failure of photographic film). Similarly, too weak or too brief an exposure of molecule B to molecule A will result in the failure of A to trigger any change in B, hence in C, and so on. Therefore, K alone is inadequate to characterize an interaction.

There are many proteins interacting in a fashion intermediate between the two extremes of transient and permanent (e.g., transcription factors that must gain a subunit in order to be able to actively bind to a promoter site).

Finally, in these preliminary remarks we recall the evolutionary constraints imposed on change: A mutation enhancing the efficiency of an enzyme may be unacceptable because of adverse changes to its quinary structure.

In the remainder of this chapter we consider the basic types of intermolecular interactions, experimental techniques for determining interactions *in vivo* and *in vitro*, and some notions about the network structure of the interactome, including its dynamical aspects.

---

[11] The dehydron (Sect. 11.5.2) is an underwrapped (i.e., underdesolvated) hydrogen bond and is a key determinant of protein affinity.

[12] See, e.g., Ramsden (1984, 1986).

16.3   Intermolecular Interactions

The simplest, and least specific, interaction is hard-body exclusion. Atoms cannot interpenetrate due to the Born repulsion. The situation is slightly more complicated for macromolecules of irregular shape (i.e., with protrusions and reëntrant hollows); they may be modelled as spheres with effective radii, in which case some interpenetration may be possible, in effect.

The Lifshitz–van der Waals force is nearly always weakly attractive, but since it operates fairly indiscriminately, not only between macromolecules but also between them and small solvent molecules, it is of little importance in conferring specificity of interaction.

Most macromolecules are ionized at cytoplasmic pH, due to dissociation (from –COOH) or addition (to $-NH_2$) of a proton, but the charge is usually effectively screened in the cytoplasmic environment, such that the characteristic distance (the Debye length) of the electrostatic interaction between charged bodies may be reduced to a fraction of a nanometre. Hence, it is mainly important for short-range steering prior to docking.

Hydrogen-bonds (H-bonds or HB) have already been encountered (Sects. 11.2, 11.3 and 11.5, etc.). A chemical group can be either an HB-donor or an HB-acceptor. Potentiated by water, this interaction can have a considerable range in typical biological milieux—out to tens of nanometres. It is the dominant interparticle interaction in biological systems.[13]

"Hydrophobic effects" or "forces" are also a manifestation of hydrogen-bonding in the presence of water, which can effectively compete for intermolecular H-bonds. The wrapping of dehydrons by appropriate apolar residues is a key contributor to protein–protein affinity.

It may be useful to think of the interactions between macromolecules in a cell as analogous to those between people at a party. It is clear that everyone is subject to hard-body exclusion. Likewise, one may feel a weak (nonspecific) attraction for everyone—misanthropes would presumably not have bothered to come. This is sufficient to allow one to fleetingly spend time exchanging a few words with a good many people, among whom there will be a few with strong mutual interest and a longer conversation will ensue. Once such mutual attraction is apparent, the conversation may deepen further, and so on. This is very like the temporal awareness shown by interacting macromolecules capable of existing in multiple states.

---

[13] Hydrogen-bonding is a special example of Lewis acid–base (AB) or electron donor–acceptor (da) interactions.

Time-Dependent Rate "Constants"

Even a two-state molecule can display temporal awareness. Consider the reaction between a receptor R that can exist in either of two states and a ligand L:

$$R_1 + L \underset{k_{d_1}}{\overset{k_a}{\rightleftharpoons}} R_1 L \, , \tag{16.8}$$

$$R_1 L \underset{k_{d_2}}{\overset{k_s}{\rightleftharpoons}} R_2 L \, ; \tag{16.9}$$

the interpretation of this would be that after initial binding, the receptor changes its conformation into that of state 2, in which the ligand is much more tightly bound. The probability of R and L remaining together can be described by a memory function, in which the amount $\nu(t)$ of associated protein is represented by the integral

$$\nu(t) = k_a \int_0^t \phi(t_1) Q(t, t_1) \, dt_1 \, , \tag{16.10}$$

where ϕ is the fraction of unoccupied binding sites. The memory kernel Q denotes the fraction of A bound at epoch t_1 that remains adsorbed at epoch t. Often, Q simply depends on the difference $t - t_1$. If dissociation is a simple first-order (Poisson) process, as is the case if the associated partners each only have a single state, then $Q(t) = \exp(-k_d t)$ and there is no memory, but in general the dissociation rate coefficient is time-dependent and can be obtained from the quotient

$$k_d(t) = \frac{\int_0^t \phi(t_1) Q'(t, t_1) \, dt_1}{\int_0^t \phi(t_1) Q(t, t_1) \, dt_1} \, , \tag{16.11}$$

where Q' is the derivative of the memory function with respect to time. A necessary condition for the system to reach equilibrium is

$$\lim_{t \to \infty} Q(t) = 0 \, . \tag{16.12}$$

Problem. Derive the memory function for the system described by the reactions (16.8). *Hint*: Use Laplace transforms.

Specificity

From the above considerations it follows that specificity of interaction is mainly influenced by geometry (due to hard-body exclusion), the pattern of complementary arrangements of HB-donors and HB-acceptors (for which an excellent example is the base-pairing in DNA and RNA (Figs. 11.3 and 11.5) and the pattern of complementary arrangements of dehydrons and apolar residues on the two associating partners.[14]

Thus, specificity of interaction (synonymous with "molecular recognition") is a kind of pattern recognition (cf. Sect. 8.2), germane to sequence matching. Clearly,

---

[14] See Ramsden (2000).

the more features that are included in the matching problem, the more discriminating the interaction will be.

Nonspecific Interactions

Most biological interactions show no discontinuity of affinity with some parameter characterizing the identity of one of the binding partners, or their joint identity, although the relation may be nonlinear. Hence in most cases the difference between specific and nonspecific interactions is quantitative, not qualitative. Even nucleotides can pair with the wrong bases, albeit with much smaller affinity.[15] In many cases, such as the association of transcription factors with promoter sites, weak nonspecific binding to any DNA sequence allows early association of the protein with the nucleic acid, whereupon the search for the promoter sequence becomes a random walk in one dimension rather than three, which enormously accelerates the finding process.[16] It should be emphasized that nonspecific binding is an essential precursor to specific binding. The scheme (16.8) applies, in which case the difference in states 1 and 2 might merely be one of orientation.

Cooperative Binding

Consider again reaction (16.5) with A representing a ligand binding to an unoccupied site on a receptor (B). Suppose that the ligand-receptor complex C has changed properties that allow it to undergo further, previously inaccessible reactions (e.g., binding to a DNA promoter sequence). The rôle of A is to switch B from one of its stable conformational states to another. The approximate equality of the intramolecular, molecule–solvent, and A–B binding energies is an essential feature of such biological switching reactions. An equilibrium binding constant K_0 is defined according to the law of mass action (16.6). If there are n independent binding sites per receptor, conservation of mass dictates that $b = nb_0 - c$, where b_0 is the total concentration of B, and the binding ratio $r = c/b_0$ (number of bound ligands per biopolymer) becomes

$$r = \frac{nK_0a}{1 + K_0a} \ . \tag{16.13}$$

Suppose now that the sites are not independent but that the addition of a second (and subsequent) ligand next to a previously bound one (characterized by an equilibrium constant K_1) is easier than the addition of the first ligand. In the case of a linear receptor B, the problem is formally equivalent to the one-dimensional Ising model of ferromagnetism, and neglecting end effects, one has

$$r = \frac{n}{2}\left(1 - \frac{1 - K_0a}{[(1 - K_0a)^2 + 4K_0a/q]^{1/2}}\right), \tag{16.14}$$

---

[15] See, e.g., Kornyshev and Leikin (2001).
[16] E.g. Ramsden and Dreier (1996) see Ramsden and Grätzel (1986) for a nonbiological example of the effect of dimensional reduction from 3 to 2.

where the degree of cooperativity q is determined by the ratio of the equilibrium constants, $q = K_1/K_0$. For $q > 1$ this yields a sigmoidal binding isotherm. If $q < 1$, then binding is anticooperative, as, for example, when an electrically charged particle adsorbs at an initially neutral surface; the accumulated charge repels subsequent arrivals and makes their incorporation more difficult.

Sustained Activation

Effective stimulation in the immune system often depends on a sustained surface reaction. When a ligand (antigen) present at the surface of an antigen-presenting cell (APC) is bound by a T-lymphocyte (TL) (see Sect. 10.5), binding triggers a conformational change in the receptor protein to which the antigen is fixed, which initiates further processes within the APC, resulting in the synthesis of more receptors, and so on. This sustained activation can be accomplished with a few, or even only one TL, provided that the affinity is not too high: The TL binds, triggers one receptor, then dissociates and binds anew to a nearby untriggered receptor (successive binding attempts in solution are highly correlated). This "serial triggering" can formally be described by

$$L + R \rightarrow R_L^* \tag{16.15}$$

(with rate coefficient k_a), where the starred R denotes an activated receptor and

$$R_L^* \rightleftharpoons R^* + L \tag{16.16}$$

with rate coefficient k_d for dissociation of the ligand L from the activated receptor, and the same rate coefficient k_a for reassociation of the ligand with an already activated receptor. The rate of activation (triggering) is $-dr/dt = -k_a r l$, solvable by noting that $dl/dt = -k_a(r + r^*) + k_d r_L^*$. One obtains

$$l(t) = \frac{k_a \tau}{1 - Y e^{-t/\tau}} + \frac{k_a(l_0 - r_0) - k_d - 1/\tau}{2k_a}, \tag{16.17}$$

where $\tau = \{4 l_0 k_a k_d + [k_a(l_0 - r_0) - k_d]^2\}^{-1/2}$ and $Y = (k_d + k_a[l_0 + r_0] - 1/\tau)/(k_d + k_a[l_0 + r_0] + 1/\tau)$, subscripts 0 denoting the initial concentrations of R and L, and the temporal evolution of the activated form is then found from

$$r(t) = r_0 \exp\left[\ln\left(\frac{1 - Y e^{-t/\tau}}{1 - Y} \right) - \frac{t}{\tau} \right]. \tag{16.18}$$

16.4 *In Vivo* Experimental Methods

Several methods have been developed involving manipulations on living cells. Although sometimes called *in vivo*, they cannot be called noninvasive. The cell is assaulted quite violently: Either it is given unnatural, but not lethal reagents, or it is killed and swiftly analysed before decay sets it, the interactions present at the moment of death being assumed to remain until they have been measured.

16.4.1 The Yeast Two-Hybrid Assay

Suppose that it is desired to investigate whether protein A interacts with protein B. The general concept behind this type of assay is to link A to another protein C, and B to a fourth protein D. C and D are chosen such that if they are complexed together (via the association of A and B), they can activate some other process (e.g., gene expression) in yeast. In that case, C could be the DNA-binding domain of a transcription factor, and D could trigger the activation of RNA polymerase. The name "hybrid" refers to the need to make hybrid proteins (i.e., the fusion proteins A-C and B-D). If A indeed associates with B, when A-C binds to the promoter site of the reporter gene, B-D will be recruited and transcription of the reporter gene will begin. The advantage of the technique is that the interaction takes place *in vivo*.

Many variants of the basic approach can be conceived and some have been realized; for example, A could be anchored to the cell membrane, and D (to which B is fused) could activate some other physiological process if B becomes bound to the membrane.

Disadvantages of the technique include the following: the cumbersome preparations needed (i.e., making the fusion proteins by genetic engineering); the possible, or even likely, modification of the affinities of A and B for each other, and of C and D for their native binding partners, through the unnatural fusion protein constructs; and the fact that the interactions take place in the nucleus, which may not be the native environment for the A–B interaction. It is also restrictive that interactions are tested in pairs only, although this does not seem to be a problem in principle; transcription factors requiring three or more proteins to activate transcription could be used.

16.4.2 Crosslinking

The principle of this approach is to instantaneously crosslink all associated partners (protein–protein and protein–DNA) using formaldehyde while the cell is still alive. It is then lysed to release the crosslinked products, which can be identified by mass spectrometry. In the case of a protein–nucleic acid complex, the protein can be degraded with a protease, and the DNA fragments to which the protein was bound—which should correspond to transcription factor binding sites—can be identified by hybridizing to a DNA microarray.

The specific instantiation for proteins (especially transcription factors) bound to DNA is called chromatin immunoprecipitation (ChIP). In order to identify the DNA, after crosslinking and cell lysis the DNA is fragmented by sonication and selected complexes are precipitated using an appropriate antibody for the protein of interest, following which the DNA can be sequenced. In order to determine where the protein binds on the chromosome, the fragmented DNA can be exposed to an appropriate microarray (ChIP-on-chip technology).

16.4.3 Correlated Expression

The assumption behind this family of methods is that if the responses of two (or more) proteins to some disturbance are correlated, then the proteins are associated. As an example, mRNA expression is measured before and after some change in conditions; proteins showing similar changes in transcriptional response (increase or decrease, etc.—the expression profile) are inferred to be associated. Another approach is to simultaneously delete (knock out) two (or more) genes that individually are not lethal. If the multiple knockout is lethal, then it is inferred that the encoded proteins are associated.

Although these approaches, especially the first, are convenient for screening large numbers of proteins, the assumption that co-expression or functional association implies actual interaction is very unlikely to be generally warranted, and, indeed, strong experimental evidence for it is lacking.

16.4.4 Other Methods

Many other ways to identify protein complexes are possible; for example, A could be labelled with a fluorophore, and B labelled with a different fluorophore absorbing and emitting at lower wavelengths. If the cell is illuminated such that A's fluorophore is excited but the emission of B's fluorophore is observed, then it can be inferred that A and B are in sufficiently close proximity that the excitation energy is being transferred from one to the other by Förster resonance. This approach has a number of undesirable features, such as the need to label the proteins and the possibility of unfavourable alignment of the fluorophores, such that energy transfer is hindered even though A and B are indeed associated.

RNA–protein binding can be investigated by the systematic evolution of ligands by the exponential enrichment (SELEX) technique, in which candidate RNA oligomers (possibly initially random) are passed through an affinity column of the protein of interest. Retained RNA is eluted, amplified using PCR, and reapplied to the column. The cycle is repeated until most of the RNA binds, whereupon it is sequenced.

16.5   *In Vitro* Experimental Methods

Here affinities are measured outside the cell. At least one of the proteins of interest has to be isolated and purified. It can then be immobilized on a chromatographic column and the entire cell contents passed through the column. Any other proteins interacting with the target protein will be bound to the column and can be identified after elution.

A much more powerful approach, because it allows precise characterization of the kinetics of both association and dissociation, is to immobilize the purified target

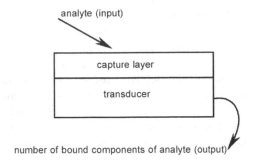

Fig. 16.1 Schematic representation of a biosensor. The thickness and structure of the capture layer, which concentrates the analyte, whose presence can then be registered by the transducer, largely determines the temporal response. The main transducer types are mechanical (cantilevers, the quartz crystal microbalance), electrical (electrodes, field-effect transistors), optoelectronic (surface plasmon resonance), and optical (planar waveguides, optical fibres). See Ramsden (1994) and Scheller and Schubert (1989) for comprehensive overviews

protein on a transducer able to respond to the presence of proteins binding to the target. The combination of capture layer and transducer is called a biosensor (Fig. 16.1).

Although this approach is formally *in vitro*, the physiological milieu can be reproduced to practically any level of detail. Indeed, as pointed out in the introduction to this chapter, the microenvironment of a subcellular compartment can be more precisely investigated than *in vivo*. Nevertheless, since each interaction is individually measured, with as much detail as is required, high throughput is only possible with massive parallelization, but because of the current expense of transducing devices, this parallelization is only practicable with protein microarrays, the penalty of which is that almost all kinetic information is lost. Hence, at present, protein microarrays and serial direct affinity measurement using biosensing devices are complementary to each other. Miniaturization of the transducers and large-scale integration of arrays of devices (comparable to the development of integrated circuit technology from individual transistors, or the development of displays in which each pixel is driven by a tiny circuit behind it) will allow the essential detailed kinetic characterization to be carried out in a massively parallel mode. Significant improvements in microarrays, allowing reliable kinetic information to be obtained from them, are also envisaged. In effect, the two approaches will converge.

16.5.1 Chromatography

Chromatography denotes an arrangement whereby one binding partner is immobilized to a solid support (the stationary phase) and the other partner is dissolved or dispersed is a liquid flowing past the solid (the mobile phase). In essence, it is like the biosensor; the difference is that binding is not measured *in situ*, but by depletion of the concentration of the mobile in the output stream. As with the biosensor, a drawback is that the immobilized protein has to be chemically modified in order to be bound to the immobile phase of the separation system. In contrast to the biosensor, the hydrodynamics within the column are complicated and chromatography is not very useful for investigating the kinetics of binding. On the other, hand there is usually an immense area of surface within the column, and the technique is therefore useful for preparative purposes.

Typically, the protein complexes are identified using mass spectrometry (examples of methods are tandem affinity purification, TAP, or high-throughput mass spectrometric protein complex identification, HMS-PCI; see Sect. 14.3).

16.5.2 Direct Affinity Measurement

As indicated in the legend to Fig. 16.1, a variety of transducers exist, the most popular being the quartz crystal microbalance (QCM), surface plasmon resonance (SPR), and optical waveguide lightmode spectroscopy (OWLS).[17] A new and even more sensitive technique is grating-coupled interferometry (GCI).[18] A great advantage of biosensors is that no labelling of the interacting proteins is required, since the transducers are highly sensitive. The order of intrinsic sensitivity is QCM < SPR < OWLS < GCI. The most sensitive method until recently (i.e., OWLS) can easily detect 1 protein per 50 μm^2 using grating couplers. Provided adequate temperature stabilization can be achieved, interferometry (i.e., optical waveguide lightmode interferometry—OWLI—for which there are various schemes, including allowing orthogonal (transverse magnetic and transverse electric) modes to propagate within the same waveguide, and dual polarization interferometry, in which the modes propagate in separate waveguides and are allowed to interfere in the far field) and the hybrid GCI can potentially achieve several orders of magnitude more sensitivity by using extended path lengths, although this may complicate the kinetic analysis of any processes being monitored.

Both QCM and SPR present a metal surface to the recreated cytoplasm, to which it can be problematical to immobilize one of the binding partners.[19] OWLS, OWLI and GCI have no such restriction since the transducer surface can be any high refractive index transparent material (titania is a popular choice). Moreover, the risk of

---

[17]See Ramsden (1994) for a comprehensive survey of all these and others.

[18]Kozma et al. (2009).

[19]A popular way to avoid the bioincompatibility of the gold or silver surface of the transducer required with SPR has been to coat it with a thick (~200 nm) layer of a biocompatible polysaccharide such as dextran, which forms a hydrogel, to which the target protein is bound. Unfortunately, this drastically changes the transport properties of the solution in the vicinity of the target (bound) protein (see Schuck 1996), which can lead to errors of up to several orders of magnitude in apparent binding constants (via a differential effect on k_a and k_d). Furthermore, such materials interact very strongly (via hydrogen bonds) with water, altering its hydrophilicity, with concomitant drastic changes to protein affinity, leading to further, possibly equally large, distortions in binding constant via its link to the free energy of interaction ($\Delta G = -RT \ln K$).

denaturing the protein by the immobilization procedure can be avoided by coating
the transducer (the optical waveguide) with a natural bilayer lipid membrane and
choosing a membrane-associated protein as the target.

For measuring the interaction, one simply causes a solution of the putative binding
protein (A) to flow over its presumed partner (B) immobilized at the transducer
surface; the binding of A to B can be recorded with very high time resolution.

The real power of this approach lies in the comprehensive characterization (i.e.,
precise determination of the number of associated proteins with good time resolution)
of the association that it can deliver. A major defect of the description built around
equation (16.5) is that the dissociation of A from B is only very rarely correctly
given by an equation of the type $d\nu/dt \sim e^{-k_d t}$, where ν is the number of associated
proteins (i.e., a pure Poisson process without memory), since most proteins remember
how long they have been associated. This is a consequence of the fact that they have
several stable states, and transitions between the states can be induced by a change
in external conditions, such as binding to another protein. The correct approach is
to consider that during a small interval of time Δt_1 at epoch t_1, a number $\Delta \nu$ of
molecules of A will be bound to B; hence,

$$\Delta \nu = k_a(\nu, t_1)\, c_A(\nu, t_1)\, \phi(\nu, t_1)\, \Delta t_1 \,, \qquad (16.19)$$

where c_A is the concentration of free (unassociated) A and ϕ is the probability that
there is room to bind (we recall that the cell is a very crowded milieu). The memory
function $Q(t, t_1)$ gives the probability that a molecule bound at epoch t_1 is still bound
at a later epoch t; hence (cf. Eq. 16.10),

$$\nu(t) = \int_0^t k_a(t_1) c_A(t_1) \phi(t_1) Q(t, t_1)\, dt_1 \,. \qquad (16.20)$$

The memory function, as well as all the other parameters in Eq. (16.20), can be
determined from the high-resolution association and dissociation kinetics.

Further advantages of the biosensor approach include the ability to study collective
and cooperative effects and to determine the precise stoichiometry of the association.

16.5.3 Protein Chips

In order to enable many interactions to be measured simultaneously, microarrays have
been developed.[20] With these arrays, the interaction of protein A with thousands of
other proteins can be studied in a single experiment, by letting A flow over the array.
Some kind of marking of A (e.g., post-reaction staining) is typically required to allow
the identification of its presence at certain sites on the array. The physical chemistry
of operation of these devices is governed by the same basic set of equations as for
the biosensor approach (Sect. 16.5), although it is not presently possible to achieve
the same sensitivity and time resolution.[21]

---

[20] Section 14.5; the immobilization of proteins without altering their conformation, and hence association characteristics, is however more difficult than for nucleic acid oligomers.

[21] See also Sect. 14.1 for limitations of the technique.

16.6   Interactions from Sequence

The principle of this approach is that gene proximity is the result of selective evolutionary pressure to associate genes that are co-regulated and, hence, possibly interacting. The motivation is to develop a method that is far less tedious and labour-intensive (and hence expensive) than the experimental techniques discussed in the preceding two sections, yet no less accurate.

Certain proteins (in a single given species) apparently consist of fused domains corresponding to individual proteins (called component proteins) in other species. The premiss of the method is that if a composite (fused or fusion) protein in one species is uniquely similiar to two-component proteins in another species, which may not necessarily be encoded by adjacent genes, those component proteins are likely to interact. "Interaction" may be either physical association or indirect functional association such as involvement in the same biochemical pathway, or co-regulation. Hence, what is inferred from this method does not exactly correspond with what is measured in the experimental methods. Nevertheless, it is an interesting attempt and one that could be developed, with more sophistication, to extract interaction data from sequence alone, which is a kind of Holy Graal for interactomics, since it is so much easier nowadays to obtain sequence data than any other kind.

16.7   Global Statistics of Interactions

The experimental difficulties are still so onerous, the uncertainties so great, and the amount of data so little that researchers have mostly been content to draw diagrams, essentially graphs, of their results, with the proteins as nodes and the associations as vertices, and leave it at that; at most, a difference in the pattern between a pair of sets of results from the same organism grown under two different conditions might be attempted. An endeavour to go beyond this first stage of representation has been made,[22] with the result (from a single dataset covering protein–protein interactions in yeast, with just under 1900 proteins and just over 2200 interactions) that the probability that a given protein interacts with k other proteins follows a power law over about one and a half orders of magnitude with an exponent $\sim - 2$. Unsurprisingly, the most heavily connected proteins were also found to be the most likely to cause lethality if knocked out.

---

[22] Jeong et al. (2001).

References

Jeong H, Mason SP, Barabási A-L, Oltvai ZN (2001) Lethality and centrality in protein networks. Nature 411:41–42

Kitano H (2002) Systems biology: a brief overview. Science 295:1662–1664

Kornyshev AA, Leikin S (2001) Sequence recognition in the pairing of DNA duplexes. Phys Rev Lett 86:3666–3669

Kozma P, Hamori A, Cottier K, Kurunczi S, Horvath R (2009) Grating coupled interferometry for optical sensing. Appl Phys B 97:5–8

Popescu O, Misevic GN (1997) Self-recognition by proteoglycans. Nature 386:231–232

Ramsden JJ (1984) The photolysis of small silver halide particles. Proc R Soc Lond A 392:427–444

Ramsden JJ (1986) Computing photographic response curves. Proc R Soc Lond A 406:27–37

Ramsden JJ (1994) Experimental methods for investigating protein adsorption kinetics at surfaces. Q Rev Biophys 27:41–105

Ramsden JJ (2000) The specificity of biomolecular particle adhesion. Colloids Surf A 173:237–249

Ramsden JJ, Dreier J (1996) Kinetics of the interaction between DNA and the type IC restriction enzyme EcoR124/3I. Biochemistry 35:3746–3753

Ramsden JJ, Grätzel M (1986) Formation and decay of methyl viologen radical cation dimers on the surface of colloidal CdS. Chem Phys Lett 132:269–272

Scheller F, Schubert F (1989) Biosensoren. Berlin: Akademie-Verlag

Schuck P (1996) Kinetics of ligand binding to receptors immobilized in a polymer matrix, as detected with an evanescent wave biosensor. I. A computer simulation of the influence of mass transport. Biophys. J. 70 1230–1249

Vohradský J (2001) Neural network model of gene expression. FASEB J 15:846–854

Vu TT, Vohradský J (2007) Nonlinear differential equation model for quantification of transcriptional regulation applied to microarray data of *Saccharomyces cerevisiae*. Nucleic Acids Res 35:279–287

The Nervous System

17

The brain is a living exemplar of an information processor *par excellence*. It seems, therefore, obvious that bioinformatics encompasses the study of the brain, although it is often considered quite separately. Because of the immense literature about the brain, this chapter will only offer a few brief insights.

The fundamental need for a brain arises because of the need of a living organism to *coordinate* its actions;[1] directive correlation implies that purpose-like behaviour requires a minimum number of causal connexions and this number is very large if activities are to be coordinated. Perfect coordination (of an activity) can be defined as implying that the activity takes account of all other activities. Even a very simple animal movement with focal condition FC (cf. Fig. 9.1) might require four muscles to be coordinated; denoting the state of excitation or inhibition of these muscles by the variables e_1, e_2, e_3 and e_4, and considering that the reaction time r of each muscle in taking account of any of the others is uniform and constant, the causal connexions for any particular epoch are shown in Fig. 17.1.

A similarly perfectly coordinated system of n muscles would require $\sim n^2 + n$ causal connexions; in practice an even greater number would be required because our variables e require both afferent and efferent connexions and further connexions would be required for the sake of adaptation to particular environmental circumstances.

Given this swift increase in complexity with size, there are evident advantages in making the connexions permanent, narrowly canalized and lacking mutual interference: a nerve system is able to satisfy these requirements. There is a further great advantage in centralizing the system as in Fig. 17.2, in which the circle represents the boundary of the nerve centre. The sixteenfold connectivity is now confined to the centre and only the fourfold afferent and fourfold efferent connectivities are required without. This is a way to achieve the greatest possible economy with respect to the length of the required communication channels.

---

[1] Sommerhoff (1950), Sect. 31.

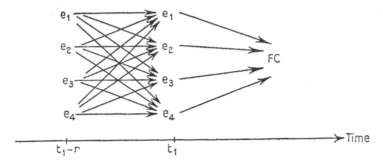

Fig. 17.1 A uniform directive correlation (Chap. 9) with focal condition FC involving four muscles whose state (of excitation or inhibition) is denoted e (see text for further explanation) (after Sommerhoff 1950; reproduced by permission of Oxford University Press [Fig. 8, p. 121])

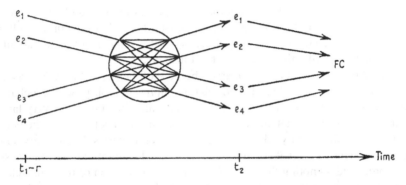

Fig. 17.2 The same scheme as in Fig. 17.1 except that the nerve connexions have been centralized (within the *circle*), showing the great economy in the overall length of the communication channels (see text for further explanation) (after Sommerhoff 1950; reproduced by permission of Oxford University Press [Fig. 9, p. 131])

17.1 The Neuron and Neural Networks

The neuron (nerve cell) is the fundamental elementary unit of the brain. It consists of a central body (the soma) with a single long and slender process growing out from it—the axon, which may, however, have many branches (collaterals). Axons impinge on receptive processes (dendrites) of other neurons or on the soma itself, the points of impingement being called synapses, which may be excitatory or inhibitory. Considering that the human brain contains about 10^{11} neurons and that each neuron may have tens of thousands of synapses, the enormous complexity of the nervous system is evident.

"Both the afferent [bringing information from distant receptors] and efferent [sending impulses to distant organs such as muscle] flow of information in the nervous system is coded in terms of the identity of the activated nerve fibres and the frequency of the impulses transmitted in them".[2] Frustratingly, the precise information that is conveyed between neurons remains unknown. Adrian established the "all or nothing" nature of the nervous impulse, and also showed that the afferent effect depends on the temporal pattern of impulses in the incoming neuron. Following receipt of inputs, via synapses, from other neurons, possible responses include firing a single pulse, firing a rapid sequence of pulses, switching between pulses and nothing ("silence").[3] The choice of response depends on the internal state of the neuron, which may also be influenced by chemicals (hormones) in its immediate environment.

Drawing on analogies with spin glasses, the structure of which is arranged to minimize energy, Hopfield (1982) proposed that the firing of neurons in networks also minimizes the energy of the network, through a process somewhat resembling epigenetic development (Fig. 10.5).

17.2 Outstanding Problems

It should be emphasized that there is no evidence that biological neural systems manifest the well-known principles of digital logic circuits. The same applies to biological memory. It may be that neural computing is not even algorithmic. In computer terms, its operations appear to be partly analogue (depending critically on the degree of excitation in the nerve fibres) and partly digital (depending critically on the identity of which fibres are activated).

One is continuously confronted with the fact that our knowledge of the brain is essentially privative. Some of the most pressing questions are as basic as how information is coded in neural activity, how memories are stored and retrieved, and what the baseline activity of the brain represents. Regarding this last question, it appears that certain (marine) animals are able to analyse noise in their sensory inputs sufficiently well to inform their hunt for prey, possibly involving stochastic resonance.

The brain is the supreme example of an object being observed—and doubtless being influenced during observation—by itself.

17.3 Artificial Neural Networks

Artificial neural nets (ANN) are inspired by the study of real networks of neurons but they have diverged from the latter and should not be considered a model of the former. They are used for computation: for example, given a set of essential features, one wishes to compute the identity of the object possessing those features.

---

[2]Sommerhoff (1974), p. 135.

[3]For details of the physicochemical mechanism of transmission of signals along the nerve fibres, see, for example, Markin et al. (1987).

An ANN typically consists of a number of individual cells ("neurons") arranged in layers and connected as follows (there are no intralayer connexions): in the first layer there are as many cells as there are input parameters for the calculation that one wishes to carry out. In the second and third layers there should be a large number of cells. Each input cell should be connected to every cell in the second layer, and every cell in the second layer should be connected to every cell in the third layer. Finally, all cells in the third layer should be connected to all the cells constituting the output layer (in some cases a single cell), which gives the result. This architecture somewhat resembles that shown in Fig. 17.2.

The connexions are channels along which information flows. The "synaptic strengths" (or conductivities) of the individual connexions increase with the amount of traffic along them. This is directly inspired by Hebb's rule for natural neural networks.

Each cell carries out a simple computation on the inputs it receives; for example it could sum the received inputs, weighted by the synaptic strengths, and output 1 if the sum exceeds some threshold, otherwise 0.

The network is trained (supervised learning) with inputs corresponding to known outputs, which fixes the synaptic strengths. It can then be used to solve practical problems. For example, one may have a set of attributes of coffee beans of unknown origin and wish to identify their provenance.

Problem. Program a general purpose computer to act as a neural network for identifying objects from their essential attributes.

References

Hopfield JJ (1982) Neural networks and physical systems with emergent collective computational abilities. Proc Natl Acad Sci USA 79:2554–2558

Markin VS, Pastushenko VF, Chizmadzhev YuA (1987) Theory of excitable media. Wiley, New York

Sommerhoff G (1950) Analytical biology. Oxford University Press, London

Sommerhoff G (1974) Logic of the living brain. Wiley, London

Metabolomics and Metabonomics

Metabolism is the ensemble of chemical transformations carried out in living tissue (Sect. 10.3); operationally it is embodied in the matter and energy fluxes through organisms. Metabolomics is defined as the measurement of the amounts (concentrations) and locations of the all the metabolites in a cell, the metabolites being the small molecules ($M_r \lesssim 1000$; e.g., glucose, cAMP,[1] GMP,[2] glutamate, etc.) transformed in the process of metabolism (i.e., mostly the substrates and products of enzymes).[3] The quantification of the amounts of expressed enzymes is, as we have seen, proteomics; metabolomics is essentially an extension of proteomics to the *activities* of the expressed enzymes, and it is of major interest to examine correlations between expression data and metabolite data.[4]

Metabonomics is a subset of metabolomics and is defined as the quantitative measurement of the multiparametric metabolic responses of living systems to pathophysiological stimuli or genetic modification, with particular emphasis on the elucidation of differences in population groups due to genetic modification, disease, and environmental (including nutritional) stress. In the numerous cases of diseases not obviously linked to genetic alteration (mutation), metabolites are the most revealing markers of disease or chronic exposure to toxins from the environment and of the effect of drugs. As far as drugs are concerned, metabonomics is effectively a subset of the investigation of the absorption, distribution, metabolism, and excretion (ADME) of drugs.

---

[1] Cyclic adenosine monophosphate.

[2] Guanosine monophosphate.

[3] The official classification of enzyme function is that of the Enzyme Commission (EC), which recognizes six main classes: 1, oxidoreductases; 2, transferases; 3, hydrolases; 4, lyases; 5, isomerases; and 6, ligases. The main class number is followed by three further numbers (separated by points), whose significance depends on the main class. For class 1, the second number denotes the substrate and the third number denotes the acceptor; whereas for class 3, the second number denotes the type of bond cleaved and the third number denotes the molecule in which that bond is embedded. For all classes, the fourth number signifies some specific feature such as a particular cofactor.

[4] These correlations are crucial for understanding the links between genome and epigenetics.

© Springer-Verlag London 2015

J. Ramsden, *Bioinformatics*, Computational Biology 21,

DOI 10.1007/978-1-4471-6702-0_18

Metabonomics usually includes not only intracellular molecules but also the components of extracellular biofluids. Of course, many such molecules have been analysed in clinical practice for centuries; the novelty of metabonomics lies above all in the vast increase of the scale of analysis; high-throughput techniques allow large numbers (hundreds) of metabolites to be analysed simultaneously and repeat measurements can be carried out in rapid succession, enabling the temporal evolution of physiological states to be monitored. The concentrations of a fairly small number of metabolites has been shown in many cases to be so well correlated with a pathological state of the organism that these metabolite concentrations could well serve as the essential variables of the organism, whose physiology is, as we may recall, primarily directed toward maintaining the essential variables within viable limits (cf. Sect. 9.4).

Metabonomics is being integrated with genomics and proteomics in order to create a new systems biology, fully cognizant of the intense interrelationships of genome, proteome, and metabolome; for example, ingestion of a toxin may trigger expression of a certain gene, which is enzymatically involved in a metabolic pathway, thereby changing it, and those changes may, in turn, influence other proteins, and hence (if some of those proteins are transcription factors or cofactors) gene expression.

18.1   Data Collection

The basic principle is the same as in genomics and proteomics: separation of the components followed by their identification. Unlike genomics and transcriptomics, metabonomics has to deal with a diverse set of metabolites, which are in some sense even more varied than proteins (which are at least all polypeptides). Typical approaches are to use chromatography to separate the components one is interested in and mass spectrometry to identify them. Alternatively, high-resolution nuclear magnetic resonance spectroscopy can be applied directly to many biofluids and even organ or tissue samples

Metabolic microarrays operate on the same principle as other kinds of microarrays (Sect. 14.1) in which large numbers of small molecules are synthesized, typically using combinatorial or other chemistry for generating high diversity. The array is then exposed to the target, whose components of interest are usually labelled (although their chemical diversity makes this more difficult than in the case of nucleic acids, for example; moreover, the small size of metabolites makes it more likely that the label chemically perturbs them). This technique can be used to answer questions such as "to which metabolite(s) does macromolecule X bind?"

Much ingenuity is currently being applied to determine spatial variations in selected metabolites. An example of a method developed for that purpose is PEB-BLES (probes encapsulated by biologically localized embedding): fluorescent dyes, entrapped inside larger cage molecules, which respond (i.e., change their fluorescence) to certain ions or molecules. Their spatial location in the cell can be mapped using fluorescence microscopy. Another example is the development of high-resolution scanning secondary ion mass spectrometry ("nanoSIMS"), whereby a focused ion beam (usually Cs^+ or O^-) is scanned across a (somewhat conducting) sample and the secondary ions released from the sample are detected mass spectrometrically with a spatial resolution of some tens of nanometres. This method is very favourable for certain metal ions, which can be detected at mole fractions of as little as 10^{-6}. If biomolecules are to be detected, it is advantageous to mark the molecule or molecules of interest by enriching them with rare but stable isotopes of their constituent atoms (e.g., ^{15}N, whose natural abundance is typically less than 1 %); the marked molecules can then easily be distinguished via the masses of their fragments in the mass spectrometer. It is usually safe to assume that the physiological effect of such marking is small.[5]

As far as whole bodies are concerned, the blood is an extremely valuable organ to analyse, since its composition sensitively depends on the state of the organism, to the extent that the blood is sometimes called "the sentinel of the body".

18.2 Data Analysis

The first task in metabonomics is typically to correlate the presence of metabolites with gene expression. One is therefore trying to correlate two datasets, each containing hundreds of points, with each other. This in essence is a problem of pattern recognition (Sect. 8.2). There are two categories of algorithms used for this task: unsupervised and supervised.

The unsupervised techniques determine whether there is any intrinsic clustering within the dataset. Initial information is given as object descriptions, but the classes to which the objects belong is not known beforehand. A widely used unsupervised technique is principal component analysis (PCA, see Sect. 8.3.2). Essentially, the original dataset is projected onto a space of lower dimension; for example, a set of metabonomic data consisting of a snapshot of the concentrations of 100 metabolites is a point in a space of 100 dimensions. One rotates the original axes to find a new axis along which there is the highest variation in the data. This axis becomes the first principal component. The second one is orthogonal to the first and has the highest residual variation (i.e., that remaining after the variation along the first axis has been taken out), the third axis is again orthogonal and has the next highest residual variation, and so on. Very often, the first two or three axes are sufficient to account for an overwhelming proportion of the variation in the original data. Since they are orthogonal, the principle components are uncorrelated (have zero covariance).

In supervised methods, the initial information is given as learning descriptions (i.e., sequences of parameter values (features) characterizing the object whose class is known beforehand).[6] The classes are nonoverlapping. During the first stage, decision

---

[5] See Voigt and Matt (2004) for some insight into this question.
[6] See, e.g., Tkemaladze (2002).

functions are elaborated, enabling new objects from a dataset to be recognized, and during the second stage, those objects are recognized. Neural networks (Sect. 17.3) are often used as supervised methods.

18.3   Metabolic Regulation

Once all of the data have been gathered and analysed, one attempts to interpret the regularities (patterns). *Simple regulation* describes the direct chemical relationship between regulatory effector molecules, together with their immediate effects, such as feedback inhibition of enzyme activity or the repression of enzyme biosynthesis. *Complex regulation* deals with specific metabolic symbols and their domains. These "symbols" are intracellular effector molecules that accumulate whenever the cell is exposed to a particular environment (cf. Table 18.1). Their domains are the metabolic processes controlled by them; for example, hormones encode a certain metabolic state; they are synthesized and secreted, circulate in the blood and, finally, are decoded into primary intracellular symbols (Sect. 18.3.2).

18.3.1 Metabolic Control Analysis

Metabolic control analysis (MCA) is the application of systems theory (Sect. 7.1) or synergetics (Sect. 7.3) to metabolism. Let $\mathbf{X} = \{x_1, x_2, \ldots, x_m\}$, where x_i is the concentration of the ith metabolite in the cell; that is, the set $\mathbf{X}$ constitutes the metabolome. These concentrations vary in both time and space. Let $\mathbf{v} = \{v_1, v_2, \ldots, v_r\}$, where v_j is the rate of the jth process. To a first approximation, each process corresponds to an enzyme. Then

$$\frac{\mathrm{d}\mathbf{X}}{\mathrm{d}t} = \mathbf{N}\mathbf{v} \,, \tag{18.1}$$

where the "stoichiometry matrix" $\mathbf{N}$ specifies how each process depends on the metabolites. Metabolic control theory (MCT) is concerned with solutions to Eq. (18.1) and their properties. The dynamical system is generally too complicated for explicit solutions to be attempted, and numerical solutions are of little use unless one knows more of the parameters (enzyme rate coefficients) and can measure more of the variables than are generally available at present. Hence, much current discussion about metabolism centres on qualitative features. Some are especially noteworthy: It is well known, from numerous documented examples, that large changes in enzyme concentration may cause negligible changes in flux through pathways of which they are a part. Metabolic networks are truly many-component systems, as discussed in Chap. 7, and, hence, the concept of feedback, so valuable in dealing with systems of just two components, is of little value in understanding metabolic networks.

Table 18.1 Some examples of metabolic coding

Condition	Symbol	Domain
Glucose deficiency	cAMP	Starvation response
N-deficiency	ppGpp	Stringent response
Redox level	NADH	DNA transcription

Problem. Write **X** and **v** in Eq. (18.1) as column matrices and **N** as an $m \times r$ matrix. Construct, solve, and discuss an explicit example with only two or three metabolites and processes.

18.3.2 The Metabolic Code

It is apparent that certain molecules mediating intracellular function (e.g., cAMP) are ubiquitous in the cell (see Table 18.1). Tomkins (1975) has pointed out that these molecules are essentially symbols encoding environmental conditions. The domain of these symbols is defined as the metabolic responses controlled by them. Note that the symbols are metabolically labile and are not chemically related to molecules promoting their accumulation. The concept applies to both within and without cells. Cells affected by a symbol may secrete a hormone, which circulates (e.g., via the blood) to another cell, where the hormone-signal is decoded—often back into the same symbol.

18.4 Metabolic Networks

Metabolism can be represented as a network in which the nodes are the enzymes and the edges connecting them are the substrates and products of the enzymes. There are two main lines of investigation in this area, which have hitherto been pursued fairly independently from one another.

The first line is centred on metabolic pathways, defined as series of consecutive enzyme-catalysed reactions producing specific products; "intermediates" in the series are defined as substances with a sole reaction producing them and a sole reaction consuming them. The complexity of the ensemble of metabolic pathways in a cell is typified by Gerhard Michal's famous chart found on the walls of biochemistry laboratories throughout the world. Current work focuses on ways of rendering this ensemble tractable; for example, a set of transformations can be decomposed into elementary flux modes. An *elementary flux mode* is a minimal set of enzymes able to operate at steady state for a selected group of transformations ("minimal" implies that inhibition of any one enzyme in the set would block the flux). A related approach is to construct linearly independent basis vectors in flux space, combinations of which express observed flux distributions. The extent to which the requirement of a steady state is realistic for living cells remains an open question. In analogy to electrical circuits, use has also been made of Kirchhoff's laws to analyse metabolic networks, especially his first law stating that the sum of all (metabolite) currents at a node is zero.

The second line is to disregard the dynamic aspects and focus on the distribution of the density of connexions between the nodes. The number of nodes of degree k appears to follow a power law distribution (i.e., the probability that a node has k edges $\sim k^{-\gamma}$).[7] Moreover, there is evidence that metabolic networks thus defined have small world properties (cf. Sect. 7.2).

Just as in the abstract networks (automata) discussed previously (Chap. 7), a major challenge in metabolomics is to understand the relationship between the physical structure (the nodes and their connecting edges) and the state structure. As the elementary demonstrations showed (cf. the discussion around Fig. 7.1), physical and state structures are only tenuously related. Much work is still needed to integrate the two approaches to metabolic networks and to further integrate metabolic networks into expression networks. Life is represented by essentially one network, in which the nodes are characterized by both their amounts and their activities, and the edges likewise.

References

Raine DJ, Norris VJ (2002) Network structure of metabolic pathways. J Biol Phys Chem 1:89–94

Tkemaladze N (2002) On the problems of an automated system of pattern recognition with learning. J Biol Phys Chem 2:80–84

Tomkins GM (1975) The metabolic code. Science 189:760–763

Voigt CC, Matt F (2004) Nitrogen stress causes unpredictable enrichments of 15N in two nectar-feeding bat species. J Exp Biol 207:1741–1748

Wagner A, Fell DA (2001) The small world inside large metabolic networks. Proc R Soc Lond B 268:1803–1810

---

[7] See Wagner and Fell (2001) or Raine and Norris (2002).

Phenomics

19

The ultimate goal, to which the study of genome and proteome lead (cf. Fig. 12.1), is to understand phenotype and phenomics is the science of the phenotype.[1] Mainly, it should be understood as the science of how to measure phenotype. In the case of static attributes (e.g., eye colour) or uniformly increasing ones (e.g., bacterial cell size under certain conditions) this is straightforward. In other cases, such as behaviour (Sect. 19.4), it is not.

19.1 Polygenic Disease

Many common diseases (e.g., asthma, diabetes and epilepsy) appear to have a genetic basis but lack the simple patterns of inheritance that would allow one to infer that they are the result of disorder in a single gene. Such *polygenic* diseases are likely much more common than single-gene diseases.

Furthermore, their incidence is known to be increasing and, although this is often attributed to environmental factors, it has been argued that it is a result of population mixing,[2] itself a corollary of globalization; the mixing tends to reintroduce susceptibility genes exogenously, which endogenously had been selected against (it can be assumed that the populations being mixed, having previously existed in different environments, have different sets of susceptibilities and resistances). Note the contrast with the benefits of mixing for diminishing the incidence of single-gene recessive diseases.[3]

---

[1] Bilder et al. (2009).

[2] Awdeh and Alper (2005) and Awdeh et al. (2006).

[3] The antithesis of polygenicity, pleiotropy (one gene affecting many traits) has been shown in at least one case to stabilize cooperation (Foster et al. 2004)—cf. Sect. 10.9.1.

© Springer-Verlag London 2015
J. Ramsden, *Bioinformatics*, Computational Biology 21,
DOI 10.1007/978-1-4471-6702-0_19

19.2   Activity-Based Protein Profiling

Although quite a few diseases can be linked to disorders in a single gene, in other cases genetic profile is a very poor predictor of susceptibility to succumbing to disease. The transcriptome also has limitations, because it does not include post-translational modifications of proteins, such as glycosylation, which can enormously change properties. To gain insight into what abnormality might be the cause of a disease, such as a metabolic disorder, it is best to directly measure the activities of key enzymes. In effect this is proteomics, but proteomics is typically only concerned with the identities and abundances of the protein repertoire. The activity-based approach preferably uses small probes interacting with enzyme active sites to give an indication of activity.[4]

One area that has been quite exhaustively investigated from this perspective is the organophosphate detoxification activity of paraoxonase (PON1), which depends on the catalytic efficiency of hydrolysis of the enzyme substrate. Although there are particular polymorphisms in the gene that change the catalytic activity of the enzyme, it has long been recognized that the determination of PON1 status requires more than genotyping:[5] both the catalytic activity of individual molecules and their abundance in serum are important.[6] Within a given population, serum PON1 activity can vary by one to two orders of magnitude; it is modulated by numerous transgenic factors, including environmental chemicals, drugs, diet and age.

19.3   Phenotype Microarrays

Microbes, with their relatively limited phenotypic repertoire, and even individual cells from multicellular organisms, are amenable to high throughput array-based assays analogous to gene microarrays (Sect. 14.1). Systems in current use are based on arrays of microwells, each well containing the cells and other necessary components.

One approach is based on cell respiration as a universal monitor of cellular activity; it is monitored colorimetrically via the reaction of NADH with a reporter dye.[7] Another approach is based on OWLS (Sect. 16.5.2) in an array format.[8] The latter is potentially much more powerful because a much more detailed view of phenotype can be obtained, including the kinetics of shape changes and of the redistribution of material within the cell.[9]

---

[4] Barglow and Cravatt (2007).
[5] Richter and Furlong (1999).
[6] Furlong (2008).
[7] Bochner et al. (2001).
[8] E.g., Orgovan et al. (2014).
[9] Ramsden and Horvath (2009).

19.4 Ethomics

The phenotype of multicellular organisms needs more sophisticated approaches than what can be achieved using a microarray (Sect. 19.3). The enormous growth in computing power has rendered feasible camera-based methods for automatically quantifying the individual and social behaviours of creatures as sophisticated as flies.[10] Such methods rely on machine-vision algorithms capable of accurately *tracking* many individual flies, and classification algorithms for the diverse *behaviours* displayed by the flies.

Even bacterial behaviour can be represented at a higher level than mere respiration: rate of growth and tumbling motions are characteristic. Any system for monitoring bacteria, however, needs to be able to contend with the short generation time, during which individuals may lose their identity and become two new entities. One approach to dealing with this problem is to use a system of interconnected compartments and ensure that each compartment contains at most one bacterium, which can then be observed unencumbered by congeners.[11]

19.5 Modelling Life

Motivations for numerically modelling living cells and organisms include the possibilities of investigating the effects of environmental factors much more rapidly and comprehensively than via actual *in vivo* experimentation, and precisely testing ideas about underlying mechanisms of biological activity. The conventional approach has been to construct a set of differential equations corresponding to all the known reactions inside a cell and in the intercellular medium and solve it numerically. It meets almost insuperable obstacles, however: not only is the system of equations, corresponding to thousands of reactions, very complicated, but many of the parameters (rate coefficients and reagent concentrations) are not reliably known.

A very different alternative approach is to emulate, rather than simulate, an organism. To this end, the cell-based virtual living organism (VLO) has been created.[12] This is a modular approach in which the exchange of information between modules plays a key rôle. Granularity plays a key role: the emulation needs to be fast enough to be useful when run on a computer but accurate enough to capture the essential features of biology. Direct simulations typically aim to model every known biochemical reaction but apart from the fact that many of the required rate coefficients and other relevant parameters are unknown, such simulations would generate vast amounts of superfluous information, obscuring the important concepts. The VLO is based on a hierarchy of the concepts of life/living, organism, organ, tissue and cell. As in a real

---

[10]E.g., Branson et al. (2009).
[11]Wakamoto et al. (2005).
[12]Bándi and Ramsden (2010, 2011).

organism, the cell plays a key rôle and all the work is done at the cellular level. In the VLO, cells give out jobs to other cells of other types and wait for the job results (which may be chemicals) when they are needed.

Models of biological systems in general, and the VLO in particular, may be useful for predicting the response of an organism to certain drugs, or the probability of creating a tumour given certain environmental conditions, and so forth. It is perhaps best viewed as a rapid prototyping tool, analogous to their very useful and already widely used counterparts in mechanical engineering.

References

Awdeh ZL, Alper CA (2005) Mendelian inheritance of polygenic diseases: a hypothetical basis for increasing incidence. Med Hypotheses 64:495–498

Awdeh ZL, Yunis EJ, Audeh MJ, Fici D, Pugliese A, Larsen CE, Alper CA (2006) A genetic explanation for the rising incidence of type 1 diabetes, a polygenic disease. J Autoimmun 27:174–181

Bándi G, Ramsden JJ (2010) Biological programming. J Biol Phys Chem 10:5–8

Bándi G, Ramsden JJ (2011) Emulating biology: the virtual living organism. J Biol Phys Chem 11:97–106

Barglow KT, Cravatt BF (2007) Activity-based protein profiling for the functional annotation of enzymes. Nat Methods 4:822–827

Bilder RM, Sabb FW, Cannon TD, London ED, Jentsch JD, Parker DS, Poldrack RA, Evans C, Freimer NB (2009) Phenomics: the systematic study of phenotypes on a genome-wide scale. Neuroscience 164:30–42

Bochner BR, Gadzinski P, Panomitros E (2001) Phenotype microarrays for high-throughput phenotypic testing and assay of gene function. Genome Res 11:1246–1255

Branson K, Robie AA, Bender J, Perona P, Dickinson MH (2009) High throughput ethomics in large groups of Drosophila. Nat Methods 6:451–457

Foster KR, Shaulsky G, Strassman JE, Queller DC, Thompson CRL (2004) Pleiotropy is a mechanism to stabilize cooperation. Nature 431:693–696

Furlong CE (2008) The Bert La Du memorial lecture paraoxonases: an historical perspective. In: Mackness B, Mackness M, Aviram M, Paragh G (eds) The paraoxonases: their role in disease, development and xenobiotic metabolism. Springer, Dordrecht, pp 3–31

Orgovan N, Peter B, Bősze Sz, Ramsden JJ, Szabó B, Horvath R (2014) Dependence of cancer cell adhesion kinetics on integrin ligand surface density measured by a high-throughput label-free resonant waveguide grating biosensor. Sci Rep 4:4034

Ramsden JJ, Horvath R (2009) Optical biosensors for cell adhesion. J Recept Signal Transduct 29:211–223

Richter RJ, Furlong CE (1999) Determination of paraoxonase (PON1) status requires more than genotyping. Pharmacogenetics 9:745–753

Wakamoto Y, Ramsden JJ, Yasuda K (2005) Single-cell growth and division dynamics showing epigenetic correlations. Analyst 130:311–317

Medical Applications

20

The question that this chapter tries to answer is, "what use is bioinformatics for medicine?" Medicine is concerned with prevention and cure of ill health and maintenance of good health. The connexion between DNA and illness once seemed clear. Well-characterized diseases such as sickle cell anaemia, known to be caused by a single point mutation in the gene coding for haemoglobin, seemed to provide solid confirmation of the "one gene, one enzyme" hypothesis.

Much of the business of bioinformatics concerns the correlation of phenotype with genotype, with the transcriptome and proteome acting as intermediaries.[1] Bioinformatics gives an unprecedented ability to scrutinize the intermediate levels and establish correlations far more extensively and in far more detail than was ever possible. This ability is revolutionizing medicine. In this spirit, one may represent the human being as a gigantic table of correlations, comprising successive columns of genes and genetic variation, protein levels, and physiological states and interactions.[2]

Medicine is mainly concerned with investigating physiological disorders, and the techniques of bioinformatics allows one to establish correlations between those disorders and variations in the genome and proteome of a patient. Medical applications of bioinformatics are mainly concerned with the investigation of deleterious genetic variation and with abnormal protein expression patterns. One can also include drug discovery as a medical application.

---

[1]Indeed, one could view the organism as a gigantic hidden Markov model (Sect. 13.5.2), in which the gene controls switching between physiological states via protein expression. Unlike the simpler models considered earlier, here the outputs could intervene in hidden layers.

[2]Since the physiological column includes entries for neurophysiological states, it might be tempting to continue the table by adding a column for the conscious experiences corresponding to the physiological and other entries. One must be careful to note, however, that conscious experience is in a different category from the entries in the columns that precede it. Hence, correlation cannot be taken to imply identity (in the same way, a quadratic equation with two roots derived by a piece of electronic hardware is embodied in the hardware, but it makes no sense to say that the hardware has two roots, despite the fact that those roots have well-defined correlates in the electronic states of the circuit components).

© Springer-Verlag London 2015
J. Ramsden, *Bioinformatics*, Computational Biology 21,
DOI 10.1007/978-1-4471-6702-0_20

20.1   The Genetic Basis of Disease

Some diseases have a clear genetic signature; for example normal individuals have about 30 repeats of the nucleotide triplet CGG, whereas patients suffering from fragile X syndrome have hundreds or thousands.

More and more data on the genotype of individuals are being gathered. Millions of single-nucleotide polymorphisms (SNPs) are now documented, and studies involving the genotyping of hundreds of SNPs in thousands of people are now feasible.[3] As pointed out earlier, most of the genetic variability across human populations can be accounted for by SNPs, and most of the SNP variation can be grouped into a small number of haplotypes.[4] This growing database is extremely useful for elucidating the genetic basis of disease, or susceptibility to disease, and hence preventive treatment for those screened routinely.

The wish to develop preventive screening implies a need for a much more rapid and inexpensive way of screening for mutations than is possible with genome sequencing. The classic method is to digest the gene with restriction enzymes and analyse the fragments separated chromatographically using Southern blotting (see footnote 1 in Chap. 14). Although direct genotyping with allele-specific hybridization is possible in simple genomes (e.g., yeast), the complexity of the human genome renders this approach less reliable. Microarrays are extensively applied to this task, as well as a related approach in which the oligonucleotides are attached to small microspheres (beads) a few micrometres in diameter. In effect, each bead corresponds to one spot on a microarray. The beads are individually tagged (e.g., using a combination of a small number of different attached fluorophores, or via the ratio of two fluorophores). Several hundred different types of beads can be mixed and discriminated at the current level of the technology. A major difficulty in the use of binding assays (hybridization) based on gene chips or beads for allele detection is the lack of complete discrimination between completely matched and slightly mismatched sequences. An alternative approach is based on the very high sequence specificity of certain enzyme reactions, such as restriction.

As well as trying to identify genes, or gene variants, responsible for disease by analysing the genome of patients, gene segments can be cloned into cells and examined for disease-like symptoms (including the pattern of expression of certain proteins). This approach is called functional cloning.

Much effort goes into understanding the correlation between gene association and disease. The rather limited success of attempts to correlate groups of SNPs with particular diseases suggests that there are many diseases enabled by combinations of two or more variant genes. The problem of correlation then acquires a combinatorial aspect and it becomes much more difficult to solve.

---

[3] These data can also be used to infer population structures (Jakobsson et al. 2008).

[4] These investigations are closely related to those of linkage disequilibrium (nonrandom association between alleles at different loci).

Many diseases have no clear genetic signature, or they depend in a complex way on genetic sequence. In cancer, for example, any relationship between gene and disease must be highly complex and has so far eluded discovery. Mutations may be important, but the changes in protein levels are equally striking. Both gene and protein chips are important here.[5]

It may well be that the impact of genetic knowledge acquired through bioinformatics will have an earlier impact on microbial infections than on intrinsic genetic disorders. It is a straightforward application of bioinformatics to design minimal microchips for the unambiguous diagnosis of a microbial infection from traces of DNA found in the blood of the patient.[6] Furthermore, the relative tractability of prokaryotic genomes will hopefully lead to an increased understanding of the nature of symbiosis, Given the ubiquity of microorganisms everywhere in our environment, symbiosis might well be considered a rather general phenomenon. The challenge is to understand multimicrobial ecosystems and how benign coexistence can sometimes suddenly become life-threatening to host metazoans.

Forensic medicine is an important branch of the medical application of genetic analysis. Repeated motifs such as variable number of tandem repeats (VNTRs) or short tandem repeats (STRs) appear to be uniquely different for each individual and, hence, can be used for identification purposes. Degradation of the DNA samples, which may have been exposed to adverse environmental influence before collection, limits the use of the longer VNTRs. The smaller STRs require PCR amplification in order to ensure that enough material is available for detection after chromatographic separation. Similar techniques are used to identify microorganisms used in biological warfare and their origin.

20.2 Cancer

Cancer, nowadays the leading cause of mortality in many developed countries, is defined as a disease involving a malignant tumour. A tumour is an abnormal lump of tissue that apparently serves no physiological purpose; it is considered to be malignant if it invades surrounding (normal) tissues or spreads to other parts of the body (a process called metastasis).

Phenotypically, cancerous cells (i.e., those constituting a tumour) are characterized by rapid and undifferentiated proliferation. Practically, the only "differentiation" that arises in a malignant tumour is angiogenesis, when the tumour is itself invaded by

---

[5]An example of the lack of a simple genetic cause of disease is illustrated by the fact that the same mutations affecting the calcium channel protein in nerve cells are observed in patients whose symptoms range from sporadic headaches to partial paralysis lasting several weeks. This is further evidence in favour of Wright's "many gene, many enzyme" hypothesis as opposed to Beadle and Tatum's "one gene, one enzyme" idea. Cf. polygenic disease (Sect. 19.1) and pleiotropy.

[6]Chumakov et al. (2005).

blood vessels, which are very necessary to ensure its continued survival and growth. Genotypically, the most characteristic feature of cancer is aneuploidy—abnormal numbers of chromosomes. Given that one of the most important factors determining reproductive isolation and, hence, speciation is chromosome mismatch, malignant tumours (cancers) may be considered to be foreign species within the host (i.e., having the status of parasites).

About 100 years ago, von Hanseman, and later Boveri, promulgated the view that aneuploidy, itself triggered by unknown causes, was the cause of cancer. Later, the idea that a few point mutations in certain genes were sufficient to upset the regulation of the cell and cause cancer received intensive scrutiny, with the genetic correlates assigned to certain stages in its progression (Table 20.1). Nevertheless, the "gene mutation theory" has a number of weaknesses, notably:[7]

1. Many chemical carcinogens are not mutagens.
2. Presumed oncomutations cannot be detected in about 50 % of cancers; conversely cells carrying presumed oncomutations are often not cancerous.
3. Presumed oncomutations are heritable, but cancers are not.
4. The probability of being afflicted by cancer increases exponentially with age. infants are essentially free of cancer, and the accumulation of the purportedly required mutations during the lifetime of a human being implies unrealistically high rates of mutation.
5. Exposure to some carcinogens results in cancer only after a very long (decades) period of latency (so-called neoplastic latency).

Cell division (mitosis), especially in eukaryotes, is an intricate affair. According to the "chromosomal theory" of cancer, carcinogens are aneuploidogens that upset the delicate and complex molecular machinery of mitosis,[8] resulting in cell division with the chromosomes unequally distributed between the two daughter cells. This results in massive genotypic aberration, equivalent to thousands of point mutations occurring in a very short time. Although many such cells are simply not viable,

Table 20.1 Stages of a cancer and some genetic correlates

Stage	Macrodescription	Microdescription
A	Dedifferentiated tissue (atavism)	Inherited mutations
B	Benign epithelial cancer	Acquired mutations: increased exposure to carcinogens from the environment
C	Adenocarcinoma	*p53* gene involved
D	Metastasis	Many mutations

---

[7] See Duesberg et al. (2005).
[8] For example, they could bind to some of the proteins, changing their affinities to the others.

presumably a few survive—indeed one may be sufficient—and form the beginning of a cancer. The abnormal karyotype appears to confer extraordinary genotypic and phenotypic instability, such that the cancer continues to evolve, developing more and more abnormal aneuploidy. Given that aneuploidy implies extra copies of some chromosomes and damage to or deletion of others, the resulting cell is likely to show phenotypically deviant behaviour, including the ability to hyperproliferate and rapidly evolve drug resistance.

20.3 Toward Automated Diagnosis

Knowledge of protein expression patterns greatly expands the knowledge of disease at the molecular level. The full power of the *pattern recognition techniques* discussed earlier (Sect. 8.2) can be brought to bear in order to elucidate the hidden mechanisms of physiological disorder. The technology of large-scale gene expression allows one to correlate gene expression patterns with disease symptoms. Microarray technology has the potential for enabling swift and comprehensive monitoring of the gene expression profile of a patient. Where correlations become well established through the accumulation of vast amounts of data, the expression profile becomes useful for diagnosis, and even for preventive treatment of a condition enhancing susceptibility to infection or allergy. One does not simply seek to correlate the bald list of expressed proteins and their abundances with disease symptoms, however: The subtleties of network structure and gene circuit topology are likely to prove more revealing as possible "causes".

The differential expression of genes in healthy and diseased tissue is usually highly revealing. For the purposes of diagnosis, each gene is characterized as a point in two-dimensional space, the two coordinates corresponding to the relative abundance of the gene product in the healthy and diseased tissue. This allows a rapid visual appraisal of expression differences.

The composition of blood is also a highly revealing diagnostic source (cf. Sect. 18.1). As well as intact peptides and other biomacromolecules, fragments of larger molecules may also be present. For their identification, mass spectrometry seems to be more immediately applicable than microarrays.

Gene chips also allow the clear and unambiguous identification of foreign DNA in a patient due to an invading microorganism, obviating the laborious work of attempting to grow the organism in culture and then identify it phenotypically.

In the future, implantable sensors are expected to be able to offer continuous monitoring of a large number of relevant physiological parameters and biomarkers (cf. Fig. 16.1). Instead of people having a biannual or even just annual blood test, hourly fluctuations could then be monitored, leading to an explosion of actimetry (activimetry) as a way of characterizing physiological state.

20.4   Drug Discovery and Testing

Whereas traditionally drugs were sought that bound to enzymes, blocking their activity, bioinformatics-driven drug discovery focuses on control points.

Intervention using drugs can take place very effectively at control points, as summarized in Table 20.2. The results of expression experiments are thus carefully scrutinized in order to identify possible control points. Once a gene or set of genes have been found to be associated with a disease, they can be cloned into cells and the encoded protein or proteins can be investigated in more detail as drug targets (functional cloning).

The proteome varies between tissues, and many different structural forms of a protein can be made by a given gene depending on cellular context and the impact of the environment on that cell. From the viewpoint of drug discovery, there are further crucial levels of detail that need to be considered, namely the way that proteins subdivide structurally into discrete domains and how these domains contain small cavities (active sites) that are considered to be the "true" targets for small-molecule drugs.

Clustering as well as other pattern recognition techniques (Sect. 18.3) can be used to identify control points in regulatory networks from proteomics and metabolomics data. DNA, RNA, and proteins are thus the significant biological entities with respect to drug development. The stages of drug development are summarized in Table 20.3. Great effort is being put into short-cutting this lengthy (and very expensive) process. For example, structural genomics can be used to predict (from the corresponding gene sequence) the three-dimensional structure of a protein suspected to be positioned at a control point. It may also be possible to compare active sites or "specificity pockets" (these regions are typically highly conserved). Toxicogenomics refers to the use of microarrays to evaluate the (adverse) effects of drugs (and toxic substances generally) across a wide range of genes, and pharmacogenomics refers to the genotyping of patients in an attempt to correlate genotype and response to a drug.

Proteins in cells do not exist in isolation. They bind to other proteins to form multiprotein structures that *inter alia* are the elements of pathways that control functions such as the responses to hormones, allergens, growth signals, and so on—things that go wrong in disease. Knowledge of the network of interactions (Sect. 16.1) is needed to understand which proteins are the best drug targets. One hopes to develop a physical map of the cell that will allow interpretation of masses of data through

Table 20.2 Stages of gene expression and their control

Stage	Description	Control (examples)
G	Genome → transcriptome (transcription)	Epigenetic regulation (networks)
T	Transcriptome → proteome (translation)	Post-translational modification
P	Proteome → dynamic system	Distributed control networks
D	Dynamic system → phenotype	Hormones
M	Metabolism	Allostery

Table 20.3 Stages of drug discovery and development

Stage	Desired outcome	Technologies involved
1. Target selection	A gene	(Functional) genomics; genotyping
2. Protein expression	A three-dimensional protein structure	Protein chemistry
3. Screening	A drug which binds	Binding studies
4. ADME	A usable drug	Interaction studies
5. Trials	An efficacious drug	Clinical trials

mining techniques and will help train predictive methods for calculating pathways and how they mesh together. Then, by homing in on the atomic details of active sites, the best candidate drug targets—probably a very small proportion of biologically valid targets—can be identified and subjected to closer scrutiny.

20.5 Nanodrugs

Nanotechnology is recognized to have sufficient potential impact on medicine for a special word, "nanomedicine", to have emerged, meaning the applications of nanotechnology to medicine. Given that medicine includes "the art of restoring and preserving health by means of remedial substances and the regulation of diet, habits, etc.", the scope of nanotechnology to intervene in medicine is large indeed. Here, however, we shall confine ourselves to describing an ingenious example of therapeutic nanoparticles (i.e., a nanodrug) involving the transmission of information.[9] The drug is actually a mixture of two different kinds of nanoparticles, "signalling" and "receiving". The rôle of the signalling particles is to target tumours. They are constructed from gold nanoparticles coated with ligands for angiogenic receptors, tumours being known to be very angiogenically active. After systemic administration these particles will tend to concentrate at the tumour due to their affinity for the angiogenic receptors. The tissue is then irradiated with an oscillating electromagnetic field, whereupon the localized nanoparticles heat up and trigger the coagulation cascade, as well as inflicting thermal damage on the tumour. The cascade essentially amplifies the information about the tumour; this information can be received by receiving particles that were also systemically introduced; these particles are equipped with coagulation-targeting peptides, but also loaded with a chemotherapeutic substance. The combination of particles enabled the chemotherapeutic dose to be increased between one and two orders of magnitude compared with a delivery system lacking the amplification–communication capability.

---

[9]von Maltzahn et al. (2012).

Problem. Carefully and critically scrutinize the von Maltzahn et al. (2012) system and subject it to a proper information-theoretic analysis.

20.6 Personalized Medicine

Given the prevalence of serious adverse drug reactions, there is much interest in identifying genetic risk factors for them, which would enable their elimination, provided that appropriate genetic screening had been carried out on the patient. A further step in that direction would be taken by organizing clinical trials of proposed new drugs such that patients are grouped according to their genetic profile. Beyond that, the development of drugs tailored to haplotype seems feasible at first sight,[10] especially with the introduction of microfluidics-based microreactors into the pharmaceutical industry, which should make reliable small-scale syntheses economically feasible.

Undertaking gene therapy evidently requires knowledge of the genome. The possibilities of direct intervention at the level of the gene have been greatly expanded by the discovery of small interfering RNA. Nevertheless, despite intensive efforts, there has been no real success in the field to date. A major problem is the difficulty of introducing the required nucleic acid material into cells from an external source.

Genome-wide association studies (GWAS) aim to scan entire (personal) genomes in order to identify genes associated with certain diseases (phenotype), especially polygenic ones (Sect. 19.1). GWAS appears to have first been proposed by Risch and Merikangas (1996). They observed that few of the numerous reports of genes or loci that might underlie complex diseases have stood up to scrutiny. They analysed linkage analysis and compared it with association analysis, using an unexceptionable model: The disease susceptibility locus has two alleles, A and a, and the genotypic relative risks (the increased chance that an individual succumbs to the disease) for genotypes aa, aA and AA are assumed to be 1, γ and γ^2, respectively. The association assumption states that the more often affected siblings share the same allele at a particular site, the more likely the site is close to the disease gene. The expected proportion of alleles shared by a pair of affected siblings is[11]

$$Y = (1+w)/(2+w) \tag{20.1}$$

where

$$w = pq(\gamma - 1)^2/(p\gamma + q)^2 \tag{20.2}$$

where p and $q = 1 - p$ are the population frequencies of A and a, respectively. If there is no linkage, $Y = 0.5$. For $p = 0.1$ and $\gamma = 4.0$, $Y =$ almost 0.6 and slightly less than 200 families would be required to make a reasonable inference of linkage. On the other hand, for the probably more realistic values of $p = 0.01$ and $\gamma = 2.0$, $Y = 0.502$ and almost 300000 families would be required, which is

---

[10]These developments are generally referred to as pharmacogenomics.

[11]See Risch and Merikangas for more about the assumptions behind these formulae.

practically unachievable. Risch and Merikangas argue that association rather than linkage tests enable an inference to be drawn from far fewer families.

Impetus in this direction came from the international HapMap project, which was based on the sequencing technology developed for the human genome project and which aimed to produce a genome-wide map of SNPs (Sect. 10.4.3). As Terwilliger and Hiekkalinna (2006) have written: "The international HapMap project was proposed in order to quantify linkage disequilibrium (LD) relationships among human DNA polymorphisms in an assortment of populations, in order to facilitate the process of selecting a minimal set of markers that can capture most of the signal from the untyped markers in a genome-wide association study. The central dogma can be summarized by the argument that if a marker is in tight LD with a polymorphism that directly impacts disease risk ... then one would be able to detect an association between the marker and the disease with sample size that was increased by a [certain] factor ... over that needed to detect the effect of functional variant directly". These authors go on to decisively refute the central dogma (of GWAS). A few years earlier, Pritchard and Cox (2002) had already written that "LD-based methods work best when there is a single susceptibility allele at any given disease locus, and generally perform very poorly if there is substantial allelic heterogeneity". Despite this, Manolio et al. (2008) were euphoric about the international HapMap project's "success", although given their affiliation with a major funding agency for the project their viewpoint may lack objectivity (the project was certainly successful at spending large sums of public money). More pertinent are remarks such as "genetic variation in chromosome ... did not improve on the discrimination or classification of predicted risk" (Paynter et al. 2009) or "treatment based on genetic testing offers no benefit compared to ... without testing" (Eckman et al.). In his paper "Considerations for genomewide association studies in Parkinson disease" (PD), Myers (2006) remarks that "Taken together, these four studies appear to provide substantial evidence that none of the SNPs originally featured as potential PD loci are convincingly replicated and that all may be false positives". It would appear that there is a great deal of evidence against the "common variants/common disease" (CV/CD) hypothesis—yet that does not prevent larger and larger studies (currently more than one million markers) being attempted. Weiss and Terwilliger were already sceptical in the year 2000 and their scepticism has been amply vindicated.

20.7 Bacterial Multiresistance

It is becoming increasingly widely perceived that one of the greatest threats to human health is the increasing ability of microbes, especially bacteria, to resist antibiotics. This resistance is a rather obvious consequence of the inept use of antibiotics,[12] but there has been little success in effectively overcoming it. One difficulty is the rapidity

---

[12]Kepler and Perelson (1998); Hermsen et al. (2012).

of the change in resistance. Analysis has shown that it occurs by the addition and rearrangement of resistance determinants and genetic mobility systems, rather than by gradual modification of the genome (Sect. 10.4.2).[13]

20.8 Toxicogenomics

It was known to Pythagoras that the broad bean, *Vicia faba*, It is poisonous to *some* people,[14] a condition known as favism and now understood to be due to genetically transmitted glucose-6-phosphate dehydrogenase (G6PD) deficiency. Such phenomena are properly the subject matter of toxicogenomics—the consequences of a particular genetic constitution for the metabolic toxicity of foods and drugs. As Tennant (2002) has pointed out, "toxicology will progressively develop from predominantly individual chemical studies into a knowledge-based science in which experimental data are compiled and computational and informatics tools will play a significant role in deriving a new understanding of toxicant-related disease". Of equal importance is the application of mRNA and protein expression technologies—that is, transcriptomics and proteomics—to study the effects of toxic substances on physiology, including metabolism, which should enable the mechanisms of toxic action to be far more effectively determined. A rarer use of the term is "the study of the response of the genome to toxic agent exposure".[15]

There are sufficient examples demonstrating that genes are not the sole determinant of the toxicities of substances; environmental factors also play a role,[16]

Problem. Evaluate the relevance of epigenetics to toxicogenomics.

20.9 Reprogramming Stem Cells

This phrase is often used to describe the remarkable discovery that by adding just four new genes to differentiated (skin) cells, after 2–3 weeks they reverted to pluripotent stem cells (induced pluripotent stem cells, iPS).[17] "Reprogramming" does not imply that cells operate like digital computers, but the discovery is of great importance, because it allows perhaps any cell to be converted into the equivalent of an embryonic stem cell, which is much more troublesome to obtain directly (from an embryo).

---

[13]Shapiro (1992).
[14]Meletis (2012).
[15]Marchant (2002).
[16]E.g., Povey (2010); see also Sect. 19.2.
[17]Takahashi and Yamanaka (2006).

20.10 Tracing Genetically Modified Ingredients in Food

Progress in the technical capabilities of plant biotechnology has resulted in the development of many transgenic plants, including many comestible ones. The motivation for introducing new genes has two primary aims: to confer resistance to pests, leading to the diminished use of chemical pesticides; and to confer resistance to herbicides, allowing their expanded and indiscriminate use to eliminate weeds. Other aims are to enhance resistance to adverse environmental conditions and to incorporate specific nutrients (e.g., vitamins) into plants that would not otherwise be considered a source for them in the human diet. In all the above cases the plants are destined for human consumption, which has given rise to some concerns about their safety. (Other uses, such as the industrial production of drugs, are not controversial.) There is not space here to delve into the controversies regarding the comestibility of genetically modified organisms (GMO); whatever the merits of the arguments on both sides, there is widespread legislative regulation of the production and distribution of GMO and, hence, a requirement for efficient and reliable methods of detecting them.

Since, by definition, GMO contain uniquely characteristic DNA sequences, their analysis is the most reliable and specific method for identifying GMO and this is, indeed, the most widespread method in use. Typically the genetic material is initially screened qualitatively, following which any samples positively identified as originating from GMO are subjected to quantitative analysis.[18]

References

Chumakov S, Belapurkar C, Putonti C et al (2005) The theoretical basis of universal identifcation systems for bacteria and viruses. J Biol Phys Chem 5:121–128

Duesberg P, Li R, Fabarius A, Hehlmann R (2005) The chromosomal basis of cancer. Cell Oncol 27:293–318

Hermsen R, Deris JB, Hwa T (2012) On the rapidity of antibiotic resistance evolution facilitated by a concentration gradient. Proc Natl Acad Sci USA 109:10775–10780

Jakobsson M, Scholz SW, Scheet P et al (2008) Genotype, haplotype and copy-number variation in worldwide human populations. Nature 451:998–1003

Kepler TB, Perelson AS (1998) Drug concentration heterogeneity facilitates the evolution of drug resistance. Proc Natl Acad Sci USA 95:11514–11519

Kutateladze T, Karseladze M, Gabriadze I, Datukishvili N (2009) Optimization of DNA-based screening methods for genetically modified organisms. J Biol Phys Chem 9:73–76

von Maltzahn G et al (2012) Nanoparticles that communicate in vivo to amplify tumour targeting. Nat Mater 10:545–552

Manolio TA, Brooks LD, Collins FS (2008) A HapMap harvest of insights into the genetics of common disease. J Clin Invest 118:1590–1605

Marchant GE (2002) Toxic genomics and toxic torts. Trends Biotechnol 20:329–332

---

[18] Kutateladze et al. (2009).

Meletis J (2012) Favism. A brief history from the "abstain from beans" of Pythagoras to the present. Arch Hellenic Med 29:258–263

Myers RH (2006) Considerations for genomewide association studies in Parkinson disease. Am J Hum Genet 78:1081–1082

Paynter NP et al (2009) Cardiovascular disease risk prediction with and without knowledge of genetic variation at chromosome 9p21.3. Ann Intern Med 150:65–72

Povey AC (2010) Gene-environmental interactions and organophosphate toxicity. Toxicology 278:294–304

Pritchard JK, Cox NJ (2002) The allelic architecture of human disease genes: common disease-common variant.. or not? Hum Mol Genet 11:2417–2423

Risch N, Merikangas K (1996) The future of genetic studies of complex human diseases. Science 273:1516–1517

Shapiro JA (1992) Natural genetic engineering in evolution. Genetica 86:99–111

Takahashi K, Yamanaka S (2006) Induction of pluripotent stem cells from mouse embryonic and adult fibroblast cultures by defined factors. Cell 126:663–676

Tennant RW (2002) The national centre for toxicogenomics: using new technologies to inform mechanistic toxicology. Environ Health Perspect 110:A8–A10

Terwilliger JD, Hiekkalinna T (2006) An utter refutation of the 'Fundamental Theorem of the HapMap'. Eur J Hum Genet 14:426–437

Ecology is undoubtedly part of biology, and information science is increasingly relevant to ecology. An especially modish activity is the analysis of time series of salient environmental parameters ("leading indicators") in order to give advance warning of imminent catastrophe.[1] Nevertheless, it is recognized that régime shifts in ecological systems can occur with no warning,[2] as in the Bak-Sneppen model of an evolving ecosystem (Sect. 10.9.2), and much rather laborious work has been undertaken to demonstrate this point. Traditionally, this topic is a branch of general systems theory (Sect. 7.1). An important rôle was played by Lotka and Volterra in the early years of the 20th century (the Lotka–Volterra model).[3] This work, in fact, long predates the emergence of general systems theory as a distinct branch of study (and may indeed have been its inspiration).

The search for "leading indicators" has by no means been abandoned. Indeed, "the potential to identify early warning signals that would allow researchers and managers to predict [e.g., the collapse of ecological communities] before they happen has therefore been an invaluable discovery" (See footnote 1). Nevertheless, the aims of this search are far from clear. If the collapse were due to external (i.e., exogenous) forcing, than indeed ecosystem "managers" might be able to do something about it (assuming that the collapse was unwanted). But if it is endogenous, then it is far from clear to what extent it could be managed.

In any case, the laborious bioinfotheoretic analysis of time series of leading indicators is scarcely necessary to give advance warning of ecosystem collapse. In most cases it is sufficiently apparent. Already Spengler (1931) could remark, "Die Mechanisierung der Welt ist in ein Stadium gefährlichsten Ueberspannung eingetreten. Das Bild der Erde mit ihren Pflanzen, Tieren und Menschen hat sich verändert. In wenigen Jahrzehnten sind die meisten grossen Wälder verschwunden, in Zeitungspapier verwandelt worden und damit Veränderungen des Klimas eingetreten, welche die Landwirtschaft ganzer Bevölkerungen bedrohen; unzählige Tierarten sind wie der

---

[1] Boettiger et al. (2013).
[2] Hastings and Wysham (2010).
[3] See the monumental review by Goel et al. (1971).

© Springer-Verlag London 2015
J. Ramsden, *Bioinformatics*, Computational Biology 21,
DOI 10.1007/978-1-4471-6702-0_21

Büffel ganz oder fast ganz ausgerottet...". Thirty years later the artist C.F. Tunni-cliffe could write, regarding the variety of wildlife in Asia, "All this [geographical] variety maintains a corresponding variety of wildlife, except in those places where humans are dominant. Thus, in India, the lion and the rhinoceros are reduced to a remnant, the Mongolian wild horse will soon be extinct, if it is not already so, and the dugong has been hunted to a shadow of its former numbers. Soon, unless man becomes suddenly more intelligent, we shall have to face fact that where he lives and works, animal life will continue to suffer, and where he is in complete control the animals must disappear completely".[4] It was not necessary to analyse time series of leading indicators to reach these conclusions.

In some cases ecosystem collapse has been managed as the result of a deliberate trade-off. An example is the fate of the Aral Sea, before 1960 the world's fourth largest inland water body, which has almost completely disappeared. It has been called the greatest man-made ecological catastrophe the world has known. To recall: the water of the Aral Sea is mainly provided by the Amu Darya (the Oxus of antiquity) and Syr Darya Rivers, and is lost by evaporation (about $60\,km^3$ per annum). The sea was the focus of a thriving fishing industry (about 40 kt per annum). These large rivers and their tributaries were already being used to a certain extent for irrigation; some canals had been constructed in the 18th century, and by 1960 about 4.5 million ha were being cultivated using irrigation, requiring about $60\,km^3$ of water per annum. At that epoch the decision was made (essentially by the central Soviet planning authorities) to vastly expand agriculture in the region. The irrigated area, and the amount of water drawn off, were roughly doubled—by 1980, in effect, the entire volume of the two big rivers was being diverted to irrigation, via a network of about 30,000 km of canals and over a hundred dams and reservoirs. Meanwhile the population also doubled, from about 14 to about 27 million during that interval. The consequences for the Aral Sea and the river deltas were already foreseen by the Moscow Hydroproject Institute in the 1960s; instead of fish (commercial fishing ceased in the early 1980s) we have the production of enormous tonnages of cotton and wheat; that was, essentially, the trade-off.[5]

---

[4]Even more stark is: "If one looks around, the world appears like an anthill where its inhabitants have lost all sense of direction. They run aimlessly about, chop each other to pieces, foul their nest, attack their young, spend tremendous energies in building artifices that are either abandoned when completed, or when maintained, cause more disruption than was visible before, and so on" (von Foerster 1972).

[5]Although one might accept the trade-off, one might criticize its implementation: much of the water taken for irrigation is lost through seepage (e.g., because of unlined canals). Excessive use of fertilizers and pesticides has led to pollution of potable water, and the now dry, salty bed of the sea is a source of aerial dust (~60 Mt/year) transported away by the wind, and the cause of widespread respiratory problems. These deleterious consequences of the transfer of water from sea to fields might have been foreseen and mitigated accordingly. A project to divert water from the great Siberian rivers to the Aral Sea was under investigation but was abandoned with the end of the Soviet Union—another kind of collapse.

It may be that modelling—albeit more sophisticated than the Lotka–Volterra equations or the Forrester–Meadows approach—informed by time series of salient data can at least present the alternatives of possible actions and prevent the inadvertent falling into disaster. Proper account of the evolving nature of the system being modelled needs to be taken.[6] That is, perhaps, the best that can be achieved; most or all ecosystems are likely examples of a Class IV cellular automaton, the evolution of which can only be explored by explicit simulation and shortcuts to the future are in principle impossible.

References

Allen PM (1998) Evolving complexity in social science. In: Altman G, Koch WA (eds) Systems—new paradigms for the human sciences. Walter de Gruyter, Berlin

Allen PM, Strathern M (2008) Complexity, stability and crises. In: Ramsden JJ, Kervalishvili PJ (eds) Complexity and security. IOS Press, Amsterdam, pp 71–92

Boettiger C, Ross N, Hastings A (2013) Early warning signals: the charted and uncharted territories. Theor Ecol 6:255–264

von Foerster H (1972) Perception of the future and the future of perception. Instruct Sci 1:31–43

Goel NS, Maitra SC, Montroll EW (1971) On the Volterra and other nonlinear models of interacting populations. Rev Mod Phys 43:231–276

Hastings A, Wysham DB (2010) Regime shifts in ecological systems can occur with no warning. Ecol Lett 13:464–472

Spengler O (1931) Der Mensch und die Technik. C.H. Beck, Munich

---

[6]Allen (1998), Allen and Strathern (2008).

The Organization of Knowledge

22

Much of biology has traditionally been concerned with the classification of objects, especially of course organisms, the best known example probably being Carl Linnaeus' *Systema Naturae*, first published in 1735. As knowledge has continued to expand, the desire to classify has also spread to bioinformatics and its objects: genes and other DNA sequences, proteins, and other molecules. As the numbers of objects stored in databases has grown, some kind of systematization has been seen as essential to aid database searches. Unfortunately, most classification almost inevitably results in distortion, and more rigid classification, the more severe the distortion. Linnaeus himself considered that his classification was to some extent artificial. The only admissible classifying arrangement of collections of objects should be that which respects the principle of maximum entropy: that arrangement should be selected, which imposes fewest assumptions upon the data.[1] Here, these issues can only be very briefly discussed; the main purpose is to alert the reader to the dangers of classification and encourage a cautious approach to their adoption. As Sommerhoff (1950) has pointed out, "Biologists have been too keen to explain things before they were able to state in exact terms what they wanted to explain", and aptly mentions Quine's remark, "that the less a science is advanced, the more does its terminology tend to rest on the uncritical assumption of mutual understanding". Ontologies (in the specific sense of footnote 3) are an obvious attempt to achieve mutual understanding, but at the price of an overly rigid structure that, given the very incomplete state of our knowledge in the field, will surely tend to hinder its further development. Just as the formation of bone requires both osteoblasts and osteoclasts, so does the growth of solid understanding require a certain conceptual fluidity, before the evidence in favour of a proposition becomes overwhelming.

---

[1]A particularly glaring example of disrespect toward this principle is to be found in the current fashion among museum curators to ceaselessly rearrange their collections in order to demonstrate some preconceived idea or another, whereas, ideally, the exhibits should be displayed in an unstructured manner, in order to allow the thoughtful visitor to draw his or her own conclusions from the raw evidence. Only in that way can new knowledge (conditional information) be generated through the perception of new, hitherto unperceived, relationships.

© Springer-Verlag London 2015
J. Ramsden, *Bioinformatics*, Computational Biology 21,
DOI 10.1007/978-1-4471-6702-0_22

Formally, classifying structures can be partitions or hierarchies. A structure s is a partition if and only if $\forall c, c' \in s, c \cap c' = \emptyset$, and it is a hierarchy if and only if $\forall i \in I, \{i\} \in s; \ \forall c, c' \in s, c \cap c' \in \{\emptyset, c, c'\}$.

Problem. Draw Venn diagrams illustrating the partition

$$\{\{a\}, \{b, c\}, \{d, e, f, g\}\},$$

and the hierarchy

$$\{\{a, b, c, d, e, f, g\}, \{d, e, f, g\}, \{b, c\}, \{e, f\}, \{a\}, \{b\}, \{c\}, \{d\}, \{e\}, \{f\}, \{g\}\}.$$

A classifying algorithm would start by constructing the classifying structure; it must then have a method (discrimination algorithm) for associating each item to be classified with a class (this is usually a pattern recognition problem; cf. Sect. 8.2), which is then applied to identify the items and place them in their classes.

22.1   Ontology

Ontology is defined as that branch of metaphysics concerned with the "nature of being". Attempts have been made to define it less metaphysically and more concretely, such as the formalization, or specification, of conceptualizations about objects in the world—including the constraints that define them individually and the relationships between them. Such formalization is held to be essential for being able to communicate with others. Hence, human languages came into being, but a problem is that they evolve: A fundamental paradox is that the desire to communicate novel, complex ideas requires individual, local innovations, which increase linguistic diversity but reduce communicability. Certain languages seem to be better than others in this regard, insofar as novel constructs can be understood by people even though they have never heard them before then.

The encapsulation of biological knowledge within database schemata almost inevitably leads to impoverishment and distortion. A good example[2] is the representation of a protein structure obtained by X-ray crystallography as an array of the three-dimensional coordinates of its constituent atoms. The raw diffraction data are refined to yield a single structure, but nearly all proteins have multiple stable structures, most of which will, however, be only slightly populated under a given set of conditions, such as those used to crystallize the protein. The protein database ignores these alternative structures.

Nevertheless, the sheer volume of data (sequences and structures) emerging from experimental molecular biology is a powerful driver for treating it ontologically in order to allow humans, and machines, to make some sense of it. Without an ontology the mass of data would be unstructured and, hence, overwhelming to the human mind, for it would be very difficult to discern meaningful paths through it.

---

[2]Pointed out by Frauenfelder (1984).

In bioinformatics, ontology typically has a more restricted definition, namely "a working model of entities and interactions".[3] These models would include a glossary of terms as a basic part. Other components of a model are generally considered to be the following (note that there has been little attempt by ontologists to define these words carefully and unambiguously): classes or categories (sets of objects); attributes or concepts, which may be either primitive (necessary conditions for membership of a class) or defined (necessary and sufficient conditions for membership); arbitrary rules (sometimes called axioms) constraining class membership, which might be considered to be part of the glossary of terms; relations (between classes or concepts), which might be either taxonomic (hierarchical) or associative; instantiations (concrete examples; i.e., individual objects); and events that change attributes, or relations, or both.

22.2 Knowledge Representation

Most obviously, knowledge representation is a medium of human expression, typically a language. In bioinformatics, the representation should be chosen to assist computation; for example, the attributes of an object being optimized using evolutionary computation (Sect. 8.1) have to be encoded in the (artificial) chromosome; it may be sufficient to represent their presence by "1" and their absence by "0", in the case of binary encoding.

Ideally, the representation should provide a guide to the organization of information—indeed knowledge might be defined as "organized (structured) information". Thus, the ontologies discussed in the previous section are an attempt to represent knowledge in this spirit. The most desirable kind of organization is that which facilitates making inductive inferences—and this will be most successfully achieved if as few preconceptions as possible are imposed on the organization.

Powerful ways of representing knowledge need not involve words, or symbolic strings, at all. Visualization (cf. Sect. 8.5) may be much more revealing than a verbal description. A particular advantage is the possibility of rearranging materials in two, rather than in one, dimension. In this regard, languages based on ideographs, most notably Chinese, would appear to be very powerful, since concepts can be rearranged on a sheet of paper and novel juxtapositions can be freely generated.

As knowledge becomes more and more complex, good examples of which are the organization of living organisms (Fig. 10.1) and their regulation (e.g., Fig. 12.1), novel ways of representing it need to be creatively explored. One approach that may

---

[3]Each different model—such as RiboWeb, EcoCyc—is typically called an "ontology"; hence, we have the Gene Ontology, the Transparent Access to Multiple Bioinformatics Information Sources (TAMBIS) Ontology (Baker et al. 1999), and so forth. If ontology is given the restricted meaning of the study of classes of objects, then an "ontology" like TAMBIS can be considered to be the product of ontological inquiry.

prove useful is to represent knowledge as probability distributions, conditional upon more or less certain facts emanating from observations or laboratory experiments; as more data becomes available, inferences can then be continuously updated in a far more systematic manner than is currently carried out today.

22.3   The Problem of Bacterial Identification

Darwin's notion of species was "a term arbitrarily given for the sake of convenience to a set of individuals closely resembling each other" (cf. the slightly more formal notion of quasispecies in sequence space: a cluster of genomes). Since bacteria predominantly proliferate asexually and can acquire new genetic material rather readily ("lateral" or "horizontal" gene transfer), the criterion of reproductive isolation that is rather helpful for defining species in metazoans is of little use. The first systematic attempt to classify bacteria dates from 1872, when Ferdinand Cohn proposed a system based on their morphology. The shape of individual bacteria can be easily seen in a (high-power) optical microscope, and colonies growing on agar plates (for example) often have characteristic morphologies themselves. Such a scheme can be readily extended to include features such as pathogenicity and characteristic biochemistry, and even characteristic habitat. The range of useful attributes depends essentially on what measuring tools are available. Thus, for example, a classification based on the compressibility of the bacterium placed between two parallel plates might also be a useful one. Gram's stain, which distinguishes between different characteristic polysaccharides coating the bacterium, is well known. This is a dichotomous classification, and a hierarchy of dichotomies should lead unerringly to the identification of a species (provided it is already known). All this knowledge has been captured in the well-known *Bergey's Manual*. Bacteria whose attributes did not match those already known would be granted the status of a new species.

The advent of molecular biology provided further vastification of the range of useful attributes. In particular, the nucleic acid sequence of the so-called 16 S ribosomal RNA (rRNA), part of the smaller subunit of the ribosome, was used by Carl Woese as a new way of classifying bacteria and, together with an assumption about the rate of mutations, could be used to construct a comprehensive phylogeny of bacteria. Bacteria seem to vary greatly in their genotypic (and phenotypic) stability, however, and any classification based on the assumption of relative stability has some limitations.[4]

---

[4]See Coenye and Vandamme (2004), and Hanage et al. (2006) for some recent discussion of the matter; Trüper (1999) has written an interesting article on prokaryotic nomenclature.

22.4 Text Mining

The literature of biology (the "bibliome")—especially research papers published in journals—has become so vast that even with the aid of a review articles that summarize many results within a few pages it is impossible for an individual to keep abreast of it, other than in some very specialized part. Text mining in the first instance merely seeks to automate the search process, by considering, above all, facts uncovered by researchers. Keyword searches, which nowadays can be extended to cover the entire text of a research paper or a book, are straightforward—an instance of string matching (pattern recognition)—but typically the results of such searches are nowadays themselves too vast to be humanly processed, and more sophisticated algorithms are required. Automated summarizing is available, based on selecting those sentences in which the most frequent information-containing words occur, but this is generally successful only where the original text is rather simply constructed. The Holy Graal in the field is the automated inference of semantic information; hence, progress depends on progress in automated natural language processing. Equations, drawings and photographs pose immense problems at present. Some protagonists even have the ambition to automatically reveal new knowledge in a text, in the sense of ideas not held by the original writer. Examples of this would be hitherto unperceived disease–gene associations.

It would certainly be of tremendous value if automatic text processing could achieve something like this level. Research papers could be automatically compared with one another, and contradictions highlighted. This would include not only contradictory facts but also facts contradicting the predictions of hypotheses. Highlighting the absence of appropriate controls, or inadequate evidence from a statistical viewpoint, would also be of great value. In principle, all of this is presently done by individual scientists reading and appraising research papers, even before they are published (through the peer-review process, which ensures (in principle) that a paper is read carefully at least once; papers not meeting acceptable standards should not (again, in principle) be accepted for publication), but the volume of papers being submitted for publication is now too large to make this method rigorously workable. Another difficulty is the already immense and still growing breadth of knowledge required to properly review many papers. One attempt to get over that problem was to start new journals dealing with small subsets of fields, in the hope that if the boundaries are sufficiently narrowly delimited, all relevant information can be taken into account. However, this is a hopeless endeavour: Knowledge is expanding too rapidly and unpredictably for it to be possible to regulate its dissemination in that way. Hence, it is increasingly likely that relevant facts are overlooked (and sometimes useful hypotheses too). Furthermore, the reviewing process is highly fragmented: It is a kind of work that is difficult to divide among different individuals; and the general trend for the number of scientists producing papers to increase exacerbates, rather than alleviates, the challenge. All that can be hoped for perhaps is that the most important results at least are properly incorporated into the edifice of reliable knowledge, but this begs the question of how to define "importance", which is often difficult to perceive in advance of what is subsequently done with the results. Another difficulty is that researchers do not always want to publish their work in what might seem to be the most appropriate journal regarding discipline: journals covering a broad range of fields and carrying a large number of advertisements tend to be disproportionately popular among scientists at present, often to the neglect of the journals published by learned societies, even those of which the authors are members.

The waters of scientific publishing have been further muddied by the emergence, and rapid growth, of open access journals. While many of these are available only online and hence much cheaper to produce than conventional printed journals, nevertheless some costs are incurred, and these are financed by *article processing charges*, which are fees charged to authors upon acceptance of a manuscript. This creates a pernicious conflict of interest for the publishers.[5] Whereas the number of subscriptions to a conventionally financed journal will depend on the quality of its content, the income of an open-access publisher is proportional to the number of papers they accept and publish. The publishers are, therefore, directly motivated to publish as many papers as possible and an easy way to achieve that is to abandon the traditions and obligations of honest and rigorous peer review.

With all of these difficulties, it is not surprising that literature mining is presently carried out in a very restricted fashion, such as merely searching for all mentions of a particular gene (and perhaps their co-occurrence with mentions of a particular disease). Whether the results of such mining are going to be useful is a moot point. There appear to be no attempts currently to weight the value of the "ore" according to some assessment of the reliability of any facts reported and assertions made. The immense difficulties still to be tackled must be weighed alongside the general growth in overall understanding (in biology) that is hopefully taking place. The edifice of reliable knowledge gradually being erected from the bricks supplied by individual laboratories allows inferences to be made at an increasingly high level, and these might well render largely superfluous endless automated reworking of the mass of facts and purported facts in the primary research literature.

One area in which it seems likely that something interesting could emerge is the search for clumps or clusters of objects (which might be words, phrases, or even whole documents) for which there is no preexisting term to describe them. Such a search might be based on a rather abstract measure of relevance (which must, of course, be judiciously chosen), along the lines suggested by Good (1962), and adumbrated in Sect. 8.3. This would be very much in the spirit of the clusters emerging when the frequencies of n-grams in DNA are examined (cf. Sect. 13.7).

If, indeed, knowledge representation moves toward probability distributions (Sect. 22.2), it would be of great value if text mining could deliver quantitative appraisals of the uncertainties of reported experimental results, which would have to include an assessment of the entire framework of the experiment (cf. Sect. 2.1.1)—that is, the structural information, as well as of the metrical information gained from the individual measurements. We seem to be rather far from achieving this

---

[5]Beall (2014).

automatically at present, but the goal merits the strongest efforts, for without such a capability, we risk being condemned to ever more fragmented knowledge, which, as a body, is increasingly shot through with internal contradictions.

22.5   The Automation of Research

Much of the laboratory work required for high-throughput genomics can be automated and carried out by laboratory robots according to a strictly executed set of instructions. In many ways this is better than carrying out the manipulations manually: the robot is likely to be able to execute its instructions more uniformly and reliably than a human experimenter. It also has advantage that a comprehensive record of the experimental conditions, as well as of the results, can be compiled automatically; this, too, may be superior to the traditional hand-written laboratory notebook, at least as far as long series of almost identical experiments are concerned. This approach also has the advantage, compared with microarray experiments (which are, of course, also robotized as a rule), that conditions individually appropriate to each experiment can be applied, avoiding the possibility of errors due to the uniform conditions applied to an entire microarray not being appropriate for some of the reactions in some of the places on the array.

A more ambitious development of automation is to automate the actual design of experiments. Given the vast scale of experiments required to elucidate gene functions and the like, this is a very necessary development. It has been realized as a robot able to measure the growth curves (defining the phenotype of a relatively simple microorganism like yeast) of selected microbial strains (distinguished by genotype) growing in defined environments.[6] The problem to which the robot has been applied is the identification of the genes for enzymes catalysing reactions thought to occur in the microbe. The robot was provided with extensive knowledge of metabolism, and software to produce hypotheses about the genes and to deduce corresponding experiments to test the hypotheses. These experiments were then executed by selecting strains from a collection given to the robot, measuring their growth curves on rich medium and then inoculating them into minimal medium to which additional metabolites, also selected by the robot, were added, after which growth curves were again measured.

Such automation is well suited to answering questions of this nature, the framework within which they are formulated being well circumscribed and carefully formulated by the investigator who actually designed the robot: essentially it functions as an extension of the brain and hands of investigator. As such, it is an extremely valuable aid and the proliferation of this technology will considerably accelerate the

---

[6]King et al. (2009).

accumulation of biological facts. The robot is certainly able to discover such facts but the (inductive) invention of knowledge remains beyond its capabilities and, perhaps, beyond the capabilities of any machine.

References

Baker PG, Goble CA, Bechhofer S et al (1999) An ontology for bioinformatics applications. Bioinformatics 15:510–520

Beall J (2014) Scholarly open-access publishing and the problem of predatory publishers. J Biol Phys Chem 14:22–24

Coenye T, Vandamme P (2004) Use of the genomic signature in bacterial classification and identification. Syst Appl Microbiol 27:175–185

Frauenfelder H (1984) From atoms to biomolecules. Helv Phys Acta 57:165–187

Good IJ (1962) Botryological speculations. In: Good IJ (ed) The scientist speculates. Heinemann, London, pp 120–132

Hanage WP, Fraser C, Spratt BG (2006) Sequences, sequence clusters and bacterial species. Phil Trans R Soc B 361:1917–1927

King RD et al (2009) The automation of science. Science 324:85–89

Sommerhoff G (1950) Analytical biology. Oxford University Press, London

Trüper HG (1999) How to name a prokaryote? FEMS Microbiol Rev 23:231–249

Bibliography

Adrian ED, Forbes A (1922) The all-or-nothing response of sensory nerve fibres. J Physiol 56:301–330

Ageno M (1992) La "Macchina" Batterica. Lombardo Editore, Rome

Ash RB (1965) Information theory. Interscience, New York

Ash RB (1998) A primer of abstract mathematics. Mathematical Association of America, Washington, DC

Audit B, Audit N, Vaillant C et al (2002) Long-range correlations between DNA bending sites: relation to the structure and dynamics of nucleosomes. J Mol Biol 316:903–918

Baharvand H (2006) Embryonic stem cells: establishment, maintenance and differentiation. In: Grier EV (ed) Embryonic stem cell research. Nova Science, Hauppauge, pp 1–63

Banzhaf W, Beslon G, Christensen S, Foster JA, Képès F, Lefort V, Miller JF, Radman M, Ramsden JJ (2006) From artificial evolution to computational evolution. Nat Rev Genet 7:729–735

Baxevanis AD, Ouellette BFF (eds) (2001) Bioinformatics, 2nd edn. Wiley, New York

Bernstein E, Allis CD (2005) RNA meets chromatin. Genes Dev 19:1635–1655

Blackburn GM, Gait MJ (1996) Nucleic acids in chemistry and biology, 2nd edn. University Press, Oxford, pp 210–221

Blumenfeld LA (1981) Problems of biological physics. Springer, Berlin

Bollobás B (1979) Graph theory. Springer, New York

Borwein P, Jörgenson L (2001) Visible structures in number theory. Am Math Mon 108:897–910

Bruinsma RF (2002) Physics of protein-DNA interaction. Physica A 313:211–237

Cacace MG, Landau EM, Ramsden JJ (1997) The Hofmeister series: salt and solvent effects on interfacial phenomena. Q Rev Biophys 30:241–278

Campbell LL (1965) Entropy as a measure. IEEE Trans Inf Theory IT-11:112–114

Cherry C (1957) On human communication. Chapman & Hall, London

Costello J, Plass C (2001) Methylation matters. J Med Genet 38:285–303

Crutchfield JP (1994) The calculi of emergence. Physica D 75:11–54

Despa F, Fernández A, Berry RS (2004) Dielectric modulation of biological water. Phys Rev Lett 93:228104

Eckman MH, Greenberg SM, Rosand J (2009) Should we test for CYP2C19 before initiating anticoagulant therapy in patients with atrial fibrillation? J Gen Intern Med 24:543–549

Edwards AWF (1972) Likelihood. University Press, Cambridge

Fell DA (1992) Metabolic control analysis: a survey of its theoretical and experimental development. Biochem J 286:313–330

© Springer-Verlag London 2015

J. Ramsden, *Bioinformatics*, Computational Biology 21,
DOI 10.1007/978-1-4471-6702-0

Felsenfeld G, Groudine M (2003) Controlling the double helix. Nature 421:448–453

Fernández A (1989) Correlation of pause sites in MDV-1 RNA replication with kinetic refolding of the growing chain. A Monte Carlo simulation of the Markov process. Eur J Biochem 182:161–163

Fickett JW (1996) The gene identification problem: an overview for developers. Comput Chem 20:103–118

Freedman DA (2009) Statistical models: theory and practice, revised edn. Cambridge University Press, Cambridge

Gell-Mann M, Lloyd S (1996) Information measures, effective complexity, and total information. Complexity 2:44–52

Gibas C, Jambeck P (2001) Developing bioinformatics computer skills. O'Reilly & Associates, Sebastopol

Gray BF (1975) Reversibility and biological machines. Nature (Lond) 253:436–437 (ibid. 257 (1975) 72)

Gull SF, Daniell GJ (1978) Image reconstruction from incomplete and noisy data. Nature (Lond) 272:686–690

Hamilton WD (1964) The genetical evolution of social behaviour. I J Theor Biol 7:1–16

Hartmann B, Lavery R (1996) DNA structural forms. Q Rev Biophys 29:309–368

Hooke R (1665) Micrographia. The Royal Society, London

Jaenisch R (1997) DNA methylation and imprinting. Trends Genet 13:323–329

James P (1997) Protein identification in the post-genome era. Q Rev Biophys 30:279–331

Kellermayer M, Ludány A, Miseta A, Koszegi T, Berta G, Bogner P, Hazlewood CF, Cameron IL, Wheatley DN (1994) Release of potassium, lipids, and proteins from nonionic detergent-treated chicken red blood cells. J Cell Physiol 159:197–204

Képès F, Vaillant C (2003) Transcription-based solenoidal model of chromosomes. Complexus 1:171–180

Kimura M (1983) The neutral theory of molecular evolution. University Press, Cambridge

Kohonen T (1988) An introduction to neural computing. Neural Netw 1:3–16

Kolmogorov AN (1965) Three approaches to the quantitative definition of information. Probl Peredachi Inf 1:3–11

Kolmogorov AN (1983) Combinatorial foundations of information theory and the calculus of probabilities. Usp Mat Nauk 38:27–36

Kolmogorov AN, Uspenskii VA (1988) Algorithms and randomness. Theor Probab Appl 32:389–412

Levich VG (1962) Physicochemical hydrodynamics. Prentice-Hall, Englewood Cliffs

Lindon JC, Nicholson JK, Holmes E, Everett JR (2000) Metabonomics: metabolic processes studied by NMR spectroscopy of biofluids. Concepts Magn Reson 12:289–320

Mackay DM (1960) Operational aspects of intellect. In: Mechanization of thought processes, Proc NPL symposium. HMSO, London, pp 37–73

Manghani S, Ramsden JJ (2003) The efficiency of chemical detectors. J Biol Phys Chem 3:11–17

Martin S, Zhang Zh, Martino A, Faulon J-L (2007) Boolean dynamics of genetic regulatory networks inferred from microarray time series data. Bioinformatics 23:866–874

McConkey EH (1982) Molecular evolution, intracellular organization, and the quinary structure of proteins. Proc Natl Acad Sci USA 79:3236–3240

McIntosh JR, Molodtsov MI, Ataullakhanov FI (2012) Biophysics of mitosis. Q Rev Biophys 45:147–207

Merton RK (1957) Priorities in scientific discovery: a chapter in the sociology of science. Am Soc Rev 22:635–659

Nicholls JG (1987) The search for connections: studies of regeneration in the nervous system of the leech. Sinauer Associates, Sunderland

Pollack GH (2001) Cells, gels and the engines of life. Ebner, Seattle

Price GR (1970) Selection and covariance. Nature (Lond) 227:520–521

Ramsden JJ (1998) Kinetics of protein adsorption. In: Malmsten M (ed) Biopolymers at interfaces. Dekker, New York, pp 321–361

Rényi A (1970) Probability theory. Akadémiai Kiadó, Budapest

Romanovsky JM, Stepanova NV, Chernavsky DS (1974) Kinetische Modelle in der Biophysik. Gustav Fischer, Jena

Ruzhentsev, Ye V (1964) The problem of transition in palaeontology. Int Geol Rev 6:2204–2213

Sanger F (1981) Determination of nucleotide sequences in DNA. Biosci Rep 1:3–18

Savageau MA (1974) Comparison of classical and autogenous systems of regulation in inducible operons. Nature (Lond) 252:546–549

Shannon CE, Weaver W (1949) The mathematical theory of communication. University of Illinois Press, Urbana

Shapiro JA (2005) Thinking about evolution in terms of cellular computing. Nat Comput 4:297–304

Shaw R (1981) Strange attractors, chaotic behaviour, and information flow. Z Naturforsch 36a:80–112

Sheldrake AR (1974) The ageing, growth and death of cells. Nature (Lond) 215:381–385

Smolen P, Baxter DA, Byrne JH (2000) Mathematical modeling of gene networks. Neuron 26:567–580

Stearns SC (1989) The evolutionary significance of phenotypic plasticity. BioScience 39:436–445

Stent G (1975) Explicit and implicit semantic content of the genetic information. The centrality of science and absolute values, 4th international conference on the unity of the sciences, vol 1. International Cultural Foundation, New York, pp 261–277

Symons MCR (1981) Water structure and reactivity. Acc Chem Res 14:179–187

Thompson TM (1983) From error-correcting codes through sphere packings to simple groups. Mathematical Association of America, Washington, DC

VanBogelen RA, Greis KD, Blumenthal RM et al (1999) Mapping regulatory networks in microbial cells. Trends Microbiol 7:320–327

Varki A et al (eds) (2009) Essentials of glycobiology, 2nd edn. Cold Spring Harbor Laboratory Press, Cold Spring Harbor

Watson JD, Crick FHC (1953) Molecular structure of nucleic acids. Nature (Lond) 171:737–738

Weiss KM, Terwilliger JD (2000) How many diseases does it take to map a gene with SNPs? Nat Genet 26:151–156

Woese CR, Olsen GJ, Ibba M, Soll D (2000) Aminoacyl-tRNA synthetases, the genetic code, and the evolutionary process. Microbiol Mol Biol Rev 64:202–236

Wright S (1982) Character change, speciation and the higher taxa. Evolution 36:427–443

Yockey HP (1958) Symposium on information theory in biology. Pergamon Press, New York

Index

Printed in the United States
By Bookmasters